THE U.S. SURVIVAL GUIDE

POCKET EDITION

NEW, IMPROVED & REMASTERED

EDITED BY
RICK CARLILE

POCKET/TRAVEL EDITION

CARLILE MEDIA

www.CARLILE.MEDIA

The US Army Survival Guide — Pocket Edition
New, Improved and Remastered

US Army
Edited by Rick Carlile
Illustrated by Carlile Media

For information purposes only.
Any action you take is at your own risk.

First published 2020 by Carlile Military Library. Carlile Military Library is an imprint of Carlile Media (a division of Creadyne Developments LLC, Las Vegas, Nevada). Carlile Military Library, Carlile Media, and their associated logos and devices are trademarks. The appearance of U.S. Department of Defense (DoD) visual information does not imply or constitute DoD endorsement.

Published in the United States of America
ISBN-13: 978-1-949117-17-2
ISBN-10: 1949117170

CARLILE
MILITARY LIBRARY

www.**CARLILE.MEDIA**
THE **ACTION** PUBLISHERS

TABLE OF CONTENTS

PREFACE **18**

INTRODUCTION **19**

OVERVIEW **20**

PERSONNEL RECOVERY 20

SURVIVAL 20

SURVIVAL MEDICINE 21

SUSTENANCE 21

PROTECTION 22

NAVIGATION 22

PSYCHOLOGY OF SURVIVAL 22

SURVIVAL STRESSORS 23

PREPARING YOURSELF 24

BE REALISTIC 24

ADOPT A POSITIVE ATTITUDE 25

TRAIN AND PREPARE 25

SURVIVAL PATTERN 25

SURVIVAL MEDICINE **28**

SURVIVAL MEDICINE VERSUS TRADITIONAL
MEDICINE 28

LIFESAVING STEPS 31

TREATING BLEEDING 32

TREATING HYPOXIA 33

DIRECT PRESSURE 35

ELEVATION 38
PRESSURE POINTS 38
DIGITAL LIGATION 40
ASSESS FOR BREATHING AND CHEST INJURIES . 40
TREATING BURNS 41
TREATING HEAD INJURY 43
TREATING SHOCK 45
ILLNESS, INFECTION, SOFT TISSUE TRAUMA 47
TREATING INFECTION 55
    BASIC TREATMENT 55
    OPEN TREATMENT 56
    GAPING WOUNDS 56
    INFECTIONS IN WOUNDS 56
    USE OF MAGGOTS 57
    DEBRIDEMENT 58
    SUTURE 58
    MEDICINAL PLANTS 58

SURVIVAL MEDICINE PROPERTIES 59
TANNIN / TANNIC ACID 59
    PREPARATION 60
    TREATMENT 60
SALICIN / SALICYLIC ACID 60
    PREPARATION 61
    TREATMENT 61
PLANTAIN / WOUNDWORT / COMMON YARROW 61
    PREPARATION 62
    TREATMENT 62
TANSY / WILD CARROT / QUEEN ANNE'S LACE .. 62
    PREPARATION 63
    TREATMENT 63
YARROW / CATTAIL / WILD ONION / GARLIC /
CHICKWEED / BURDOCK 63
    PREPARATION 63
    TREATMENT 63

POISONOUS PLANTS 64
RULES FOR AVOIDING POISONOUS PLANTS 65
    AVOIDING CONTACT DERMATITIS 65
    AVOIDING INGESTION POISONING 66

TREATING INJURIES, BITES AND STINGS,
POISONING, AND OTHER CONDITIONS 67
SOFT TISSUE TRAUMA 67
    BLISTERS 67

BOILS 67
FUNGAL INFECTIONS 68
FACIAL INJURY 68
EYE INJURY 69
BONE AND JOINT TRAUMA 69
SPRAINS 70
STRAINS 70
FRACTURES AND DISLOCATIONS 70
INSECT BITES AND STINGS 73
SNAKEBITES 74
ENVIRONMENTAL INJURY 76
HIGH ALTITUDE SICKNESS 76
HIGH-ALTITUDE PULMONARY EDEMA 76
HEAT INJURIES 77
COLD INJURIES 77
CHILBLAINS AND FROSTBITE 79
IMMERSION FOOT 79
HYPOTHERMIA 79
DEHYDRATION 80
INGESTION POISONING 81

PERSONAL HYGIENE AND SANITATION 82
AVOIDING ILLNESS 82
INTESTINAL PARASITES 83

WATER **85**
HYDRATION CONSIDERATIONS 85
WATER SOURCES AND INDICATORS 86
PROCURING WATER 86
RAIN, SNOW, AND ICE 89
DEW 91
WATER FROM PLANTS 91
GREEN BAMBOO 91
WATER VINES 92
BANANA PLANTS 93
WATER TREES 94
COCONUT WATER 95
VEGETATION BAG STILL 96
TRANSPIRATION BAG STILL 97
BELOW-GROUND SOLAR STILL 98
BEACH WELLS AND SALTWATER 100
SEEPAGE WELLS AND SUB-SURFACE WATER ... 100
PREPARING WATER 101
DRINKING WATER HAZARDS 101

FILTRATION 102
PURIFICATION 103
BOILING METHOD 104
CHEMICAL PURIFICATION 104
   TYPES OF CHEMICAL PURIFICATION 104
   COMMERCIAL FILTERS 105

FOOD 106

FOOD CONSIDERATIONS 106
NUTRITION 107
   CARBOHYDRATES 107
   FATS 107
   PROTEINS 107
   VITAMINS 108
   MINERALS 108
FOOD AVAILABILITY 108
   PLANTS 108
   ANIMALS 109
BASIC FOOD PREPARATION 109

BASIC COOKING AND PRESERVATION METHODS 111
LEACHING 111
BOILING 111
   PARBOILING 112
   STEAMING 112
   STONE BOILING 112
BAKING 113
   REFLECTOR OVEN 113
   BAKING WITH LEAVES 113
ROASTING 113
FRYING 114

PRESERVING FOOD 114
SUN DRYING 115
SMOKING 115

PLANTS 115
PLANT IDENTIFICATION 119
PLANT EDIBILITY TESTING 122
GLOBALLY-COMMON EDIBLE PLANTS 125
SEAWEED 126
PREPARING PLANTS FOR CONSUMPTION 127

MAMMALS 127
HUNTING 128

TRAPPING AND SNARES 130
USE OF BAIT 132
CONSTRUCTION OF TRAPS AND SNARES 132
TRIGGERS FOR TRAPS AND SNARES 132

KILLING DEVICES 144
RABBIT STICK 144
SPEAR 144
SLING 144

BUTCHERING MAMMALS 145
SKINNING MAMMALS 145
BIG GAME 145
SMALL GAME 148

BIRDS 148

INSECTS 151

REPTILES 152

AMPHIBIANS 153

FISH 154
FISHING DEVICES 156
WOODEN HOOK 157
GORGE OR SKEWER 157
STAKEOUT 158
GILL NET 158
FISH TRAPS 160
CHOP FISHING 162
POISON 162
PREPARING FISH FOR CONSUMPTION 165

MOLLUSKS 165

CRUSTACEANS 167

FIRE 168
FIRE BASICS 168
FIRE BURNING PRINCIPLES 169
SELECTING AND PREPARING A FIRE SITE 169
FIRE REFLECTOR 170
BUILDING A FIRE 171
TINDER 173
KINDLING 174
FUEL 175
FIRE LAYS 177
TEPEE 177
LEAN-TO 177

CROSS-DITCH 177
PYRAMID 178
LOG CABIN 178
LONG FIRE 178
STAR FIRE 178

FIRE STARTING TECHNIQUES 179
MODERN METHODS OF FIRE BUILDING 180
MATCHES 180
LIGHTERS 180
METAL MATCH 181
BATTERIES 182
CONVEX LENS 182
FLASHLIGHT REFLECTOR 183
PRIMITIVE METHODS OF FIRE BUILDING 183
FLINT AND STEEL 183
FIRE PLOW 184
BOW AND DRILL 185
SOCKET 185
DRILL 185
FIRE BOARD 186
FIRE PAN 186
BOW 186
TINDER 186
OPERATING A BOW AND DRILL 186
HAND DRILL 188
FIRE BUNDLE 189

SHELTER AND CLOTHING 191
PRIMARY SHELTER 191
SHELTER CONSIDERATIONS 194
SHELTER SITE SELECTION 194
SITE PREPARATION 195
CONSTRUCTION METHODS 195
IMMEDIATE ACTION SHELTER 195
NATURALLY OCCURRING SHELTERS 196
MANMADE SHELTER CONSTRUCTION 197
PONCHO LEAN-TOS 197
PONCHO TENT 199
A-FRAME SHELTERS 201
NATURAL SHELTER CONSTRUCTION 203
SNOW CAVES 204
TREE PIT SHELTER 205
RAISED PLATFORM SHELTER 206

FIELD-EXPEDIENT LEAN-TO 207

DESERT SHELTERS 208

SWAMP BED 209

DEBRIS HUT 210

THERMAL PRINCIPLES AND INSULATION 211

TIE-OFF POINTS 212

STAKING 213

BEDDING AND GROUND INSULATION 213

MOVEMENT AND NAVIGATION 214

DECISION TO STAY OR MOVE 214

MOVEMENT CONSIDERATIONS 215

MOUNTAINOUS AND COLD MOVEMENT 216

DESERT MOVEMENT 217

JUNGLE MOVEMENT 218

WATER CROSSINGS 218

RIVERS AND STREAMS 218

RAPIDS 219

RAFTS 222

BRUSH RAFT 222

AUSTRALIAN PONCHO RAFT 223

LOG RAFT 225

FLOTATION DEVICES 226

DETERMINE CARDINAL DIRECTION 227

SHADOW TIP METHOD 227

EQUAL-SHADOW METHOD 229

WATCH METHOD 230

24-HOUR CLOCK METHOD 231

USING THE MOON 232

USING THE STARS 232

NORTHERN SKY 232

STARS 233

SOUTHERN SKY 234

IMPROVISED COMPASS 235

FLOATING NEEDLE OR LEAF COMPASS 236

OTHER MEANS OF DETERMINING DIRECTION 236

NAVIGATION METHODS 236

GLOBAL POSITIONING SYSTEM 237

DEAD RECKONING 237

TERRAIN ASSOCIATION 240

IDENTIFYING AND LOCATING SELECTED
FEATURES 241

USING HANDRAILS, CATCHING FEATURES, AND

NAVIGATIONAL ATTACK POINTS 241
HANDRAILS 241
CATCHING FEATURES 241
NAVIGATIONAL ATTACK POINTS 242
COMBINING TECHNIQUES 242
MAP ORIENTATION 242
USING A COMPASS 242
WITHOUT A COMPASS 244
TERRAIN FEATURES 245
MEASURING DISTANCES 251
GRAPHIC SCALES 251
GROUND DISTANCE—STRAIGHT LINE 251
GROUND DISTANCE—CURVED LINE 252
MEASURING AZIMUTH 253
PLOTTING A DIRECT LINE 254
MEASURING DISTANCES BY PACES 254
FIELD-EXPEDIENT GUIDELINES FOR DISTANCE TO A
LANDMARK 255

SURVIVAL EQUIPMENT 257
MAINTAINING EQUIPMENT 257
PROTECTION 257
LUBRICANTS AND GLUE 258
FIELD-EXPEDIENT WEAPONS 258
IMPROVISING 258
STAFFS 259
CLUBS 259
SIMPLE CLUB 259
WEIGHTED CLUB 259
SLING CLUB 260
EDGED WEAPONS 261
KNIVES 261
STONE 261
BONE 263
WOOD 263
METAL 263
OTHER MATERIALS 264
SPEAR BLADES 264
ARROW POINTS 265
OTHER EXPEDIENT WEAPONS 265
THROWING STICK 265
ARCHERY EQUIPMENT 265
BOLA 267

SLING 268

SLINGSHOT 268

SAP 268

CORDAGE AND LASHING 269

NATURAL CORDAGE SELECTION 269

LASHING MATERIAL 269

RUCKSACK CONSTRUCTION 270

COOKING AND EATING UTENSILS 272

BOWLS 272

FORKS, KNIVES, SPOONS AND CHOPSTICKS 272

POTS 272

IMPROVISING POTTERY 273

WATER CONTAINERS 274

WEAVING BASKETS 274

SURVIVAL KNOTS AND ROPE 276

ROPE TERMINOLOGY 276

BASIC KNOTS 278

OVERHAND KNOT 278

ROUND TURN AND TWO HALF HITCHES 279

FIGURE-EIGHT KNOT 280

KNOTS FOR JOINING ROPE 281

SQUARE KNOT 281

SINGLE SHEET BEND 282

DOUBLE SHEET BEND 283

KNOTS FOR MAKING LOOPS 283

BOWLINE 284

BOWLINE ON A BIGHT 284

FRENCH BOWLINE 285

SPEIR KNOT 286

OVERHAND KNOT FIXED LOOP 287

HITCHES 287

HALF HITCH 288

TWO HALF HITCHES 288

TIMBER HITCH 289

TIMBER HITCH AND HALF HITCH 290

CLOVE HITCH 291

SHEEPSHANK 291

PRUSIK 292

LASHINGS 293

SQUARE LASHING 293

DIAGONAL LASHING 294

SHEAR LASHING 295

ROPE CONSTRUCTION 296

NATURAL FIBER TYPES 296

MANMADE FIBER TYPES 297

CONSTRUCTION TECHNIQUES 297

TWISTING TECHNIQUE 298

BRAIDING TECHNIQUES 299

THREE-PLAIT BRAID 300

BROAD BRAID 300

WHIPPING THE ENDS 302

GLOSSARY 305

SECTION 1: ACRONYMS AND ABBREVIATIONS . 305

SECTION 2: TERMS 305

REFERENCES 306

RELATED PUBLICATIONS 306

INDEX 307

LIST OF FIGURES

SURVIVAL ACRONYM 26
JAW THRUST METHOD 34
APPLICATION OF A PRESSURE DRESSING 37
KEY BODY PRESSURE POINTS 39
TREATMENT OF SHOCK 47
BUILDING A SPLINT 72
SLINGS 73
HEAT TRANSFER 78
BAMBOO WATER CATCH (OR RAIN CATCHER) 89
WATER MACHINE 90
PROCURING WATER FROM GREEN BAMBOO 91
VINE WATER PROCUREMENT TECHNIQUE 93
WATER FROM BANANA STUMP 94
HOW TO OPEN A COCONUT 96
VEGETATION BAG STILL 97
TRANSPIRATION BAG STILL 98
BELOW-GROUND SOLAR STILL 99
BEACH WELL 100
SUB-SURFACE WATER 101
WATER FILTERING SYSTEMS 103
BOILING 112
UMBRELLA-SHAPED FLOWERS 118
LEAF MARGINS 119
LEAF SHAPES 120
LEAF ARRANGEMENTS 121
ROOT STRUCTURES 122
SIMPLE LOOP SNARE 134

APACHE FOOT SNARE 134
DRAG NOOSE SNARE 135
TWITCH-UP SNARE 136
SQUIRREL POLE NOOSE 137
NOOSING WAND 138
TREADLE SPRING SNARE 139
DEADFALL 140
PAIUTE DEADFALL 141
BOW TRAP 142
PIG SPEAR SHAFT 143
BOTTLE TRAP 144
DEER SKINNING EXAMPLE 147
MIST NET BIRD SNARE 149
OJIBWA BIRD POLE 150
CLEANING A SNAKE 152
TURTLES WITH POISONOUS FLESH 153
AMPHIBIANS 154
FISH WITH POISONOUS FLESH 156
IMPROVISED FISH HOOKS 157
STAKEOUT 158
MAKING A GILL NET 159
SETTING A GILL NET IN A STREAM 160
VARIOUS TYPES OF FISH TRAPS 161
TYPES OF SPEAR POINTS 162
FISH-POISONING PLANTS 164
EDIBLE MOLLUSKS 166
FIRE TRIANGLE 169
TYPES OF FIREWALLS 171
BASE AND BRACE 172
TINDER AND KINDLING EXAMPLES 172
FIRE LAYS 179
STARTING A FIRE WITH A BATTERY 182
USING A CONVEX LENS AS A FIRE-STARTER 183
FLINT AND STEEL FIRE-STARTER 184
FIRE PLOW 185
BOW AND DRILL FIRE-STARTER 188
HAND DRILL FIRE-STARTER 189
FIRE BUNDLE 190
TYPES OF NATURALLY OCCURRING SHELTERS 197
PONCHO LEAN-TO 199
PONCHO TENT USING OVERHANGING BRANCH 200
PONCHO TENT USING CENTER SUPPORT 200
TYPES OF MAN-MADE SHELTERS 201

A-FRAME SHELTERS 203

SNOW CAVE 205

TREE-PIT SNOW SHELTER 206

RAISED PLATFORM SHELTER 207

FIELD-EXPEDIENT LEAN-TO AND FIRE REFLECTOR 208

BELOW-GROUND DESERT SHELTER 209

SWAMP BED 210

DEBRIS HUT 211

PLACEMENT OF VENTILATION HOLES IN A SNOW CAVE ...
212

USING A POLE METHOD TO FORD STREAM 220

USING A POLE FOR MULTIPLE CROSSING 221

RIVER CROSSING USING ROPE 222

BRUSH RAFT 223

AUSTRALIAN PONCHO RAFT 225

LOG RAFT 226

TWO-LOG RAFT 227

SHADOW-TIP METHOD 228

EQUAL-SHADOW METHOD 230

WATCH METHOD 231

NORTHERN SKY 233

SOUTHERN SKY 234

MAP ORIENTATION EXAMPLES 244

ORIENTING A MAP TO TERRAIN FEATURES 245

A HILL 246

A SADDLE 246

A VALLEY 247

A RIDGE 248

A DEPRESSION 248

A DRAW 249

A SPUR 250

A CLIFF 250

GRAPHIC SCALE 251

MEASURE MAP DISTANCE—STRAIGHT LINE 252

MEASURE MAP DISTANCE—CURVED LINE 253

MEASURING AZIMUTH ON A MAP 253

PLOTTING A DIRECTION 254

WEIGHTED CLUBS 260

SLING CLUB 261

STONE KNIFE 262

BAMBOO FOR SPEARS 264

THROWING STICK 265

ARCHERY EQUIPMENT (BOW) 267

BOLA 268
MAKING CORDAGE 269
HORSESHOE PACK 271
SQUARE PACK 271
CONTAINERS FOR BOILING FOOD 273
ELEMENTS OF ROPES AND KNOTS 278
OVERHAND 279
ROUND TURN AND TWO HALF HITCHES 280
FIGURE-EIGHT KNOT 281
SQUARE KNOT 282
SINGLE SHEET BEND 283
DOUBLE SHEET BEND 283
BOWLINE KNOT 284
BOWLINE ON A BIGHT KNOT 285
FRENCH BOWLINE 286
SPEIR KNOT 287
HALF HITCH AND A TWO HALF HITCH 289
TIMBER HITCH 290
TIMBER HITCH AND HALF HITCH 290
CLOVE HITCH 291
SHEEPSHANK 292
PRUSIK 293
SQUARE LASHING 294
DIAGONAL LASHING 295
SHEAR LASHING 296
TWISTING FIBERS 299
THREE-STRAND BRAID 300
BROAD BRAID 301
FINISH BRAID 302
WHIPPING THE END OF A ROPE 304

LIST OF TABLES

SERE PROFICIENCIES 21
TYPICAL ILLNESSES, INFECTIONS, CAUSES AND
    TREATMENTS 48
MEANS TO MAKE POTABLE WATER 86
TENSILE STRENGTH OF NATURAL FIBERS 298

PREFACE

"As a soldier, you can be sent to any area of the world. It may be in a temperate, tropical, arctic, or subarctic region. You expect to have all your personal equipment and your unit members with you wherever you go. However, there is no guarantee it will be so. You could find yourself alone in a remote area—possibly enemy territory—with little or no personal gear. This manual provides information and describes basic techniques that will enable you to survive and return alive should you find yourself in such a situation."

This edition of the U.S. Army Survival Guide is based on Army Techniques Publication (ATP) 3-50.21, which in 2018 replaced the larger FM 3-05.70,[1] which in turn replaced the obsolete FM 21-76 in 2002. At nearly 700 pages in length the older FM3-05.70 is better suited to desktop use, whereas the newer version concentrates the information down to a more portable size that provides similar utility and is ideal not only for reading anywhere, but for keeping in a bug-out bag or vehicle for use in an emergency. As the old saying goes, the superior tool is the one you actually have with you when you need it.

Furthermore, this edition has been extensively edited, corrected, annotated and reformatted, and its illustrations have been completely recreated in order to improve clarity and readability, particularly under low light or stress-inducing conditions.

Rick Carlile, Editor

---

1. Also available from Carlile Media: ISBN 1547209461.

INTRODUCTION

This book sets forth the doctrine pertaining to survival in an isolated situation. It discusses the tenets of survival and the methods Soldiers, civilians, and contractors can use when surviving individually or in a group. The personnel recovery mission includes preparing Army personnel in danger of isolation while participating in any activity or mission sponsored by the United States. "Isolation" refers to the state of persons separated from their unit or in a situation where they must survive, evade, resist, or escape.

The following is a brief description of each chapter and appendix:

- **Chapter 1** provides an overview of personnel recovery and discusses survival proficiencies.
- **Chapter 2** discusses survival medicine applications.
- **Chapter 3** covers water collection methods.
- **Chapter 4** discusses food collection and preparation methods.
- **Chapter 5** focuses on fire craft for survival.
- **Chapter 6** covers constructing shelters in the field and clothing.
- **Chapter 7** discusses land navigation methods.
- **Chapter 8** covers survival, evasion and recovery equipment.
- **Appendix A** discusses ropes and knots useful for survival applications.

CHAPTER 1

OVERVIEW

Survival is the state of or fact of continuing to live or exist in spite of an isolating event. The end goal is to return to friendly forces with honor. This chapter discusses survival skills and situational understanding needed.

PERSONNEL RECOVERY

1-1. Army personnel recovery refers to the military efforts taken to prepare for and execute the recovery and reintegration of isolated personnel (FM 3-50). Personnel recovery aims to return isolated personnel to duty as well as to sustain morale, increase operational performance, collect information, and develop intelligence. Army forces work with unified action partners to recover individuals and groups who become isolated. "Isolation" refers to the state of persons separated from their unit or in a situation where they must survive, evade, resist, or escape.

1-2. All Soldiers, Department of Army civilians and supporting contractors must understand how Army survival, evasion, resistance, and escape (SERE) relates to personnel recovery. Survival, evasion, resistance, and escape is defined as the actions performed by isolated personnel designed to ensure their health, mobility, safety, and honor in anticipation of or preparation for their return to friendly control (JP 3-50). Army SERE represents actions that enable isolated individuals to survive and return to areas under friendly control as soon as possible. SERE focuses on the actions of the isolated person or group of isolated persons.

SURVIVAL

1-3. Table 1-1 identifies the four SERE actions and their related proficiencies. Survival begins when a Soldier executes their isolated Soldier guidance (ISG) or evasion plan of action (EPA). The survival proficiencies include protection, sustenance, survival medicine, and navigation. Units and individuals train on these proficiencies to improve and sustain individual task proficiency.

Table 1-1. SERE Proficiencies

Survival	Evasion
Protection.	Capture avoidance.
Sustenance.	Detection avoidance.
Survival medicine.	Recognition avoidance.
Navigation.	Communication.
	Recovery.

Resistance	Escape
Organization.	Restraint defeat.
Exploitation resistance.	Cell defeat.
Captivity communication.	Building defeat.
Health.	Installation defeat.
Honor.	

1-4. The recovery of isolated personnel and their return to friendly control is dependent upon their ability to employ relevant knowledge and skills of the survival proficiencies tailored to their operational environment.

SURVIVAL MEDICINE

1-5. Survival medicine is characterized by remote and improvised care of isolated personnel with routine or exotic physical or psychological illnesses or trauma, limited resources and labor, and delayed evacuation to definitive care. There are four fundamentals common to survival medicine: prevention, recognition, mitigation, and treatment. Assess these fundamentals during planning and preparation activities and include relevant physical and psychological factors. Additionally, they form the basis for effective organization, training, and equipping efforts by the commander and staff, unit recovery force and the individual. Planning is the art and science of understanding a situation, envisioning a desired future, and laying out effective ways of bringing that future about (ADP 5-0).

SUSTENANCE

1-6. Sustenance includes procuring, preparing, and storing adequate food and water to enable continuity of the isolated person's normal body functions and to provide strength, energy, and endurance to overcome the stress of survival. Isolated personnel must be prepared to use a variety of both native and domestic food sources, obtain water using various methods, and be able to prepare and preserve food and water throughout the duration of their isolation. Assess life-sustaining food and water needs during planning and preparation activities and integrate into organization, training, and equipping considerations by the commander and staff, unit recovery force, and the individual. Preparation are those activities by units and Soldiers to improve their ability to execute operations (ADP 5-0).

PROTECTION

1-7. Protection includes care and repair of protective clothing and equipment, the ability to build appropriate shelter and fire and the application of basic security measures. Protection needs are assessed during planning and preparation activities and integrated into organization, training, and equipping considerations by the commander and staff, unit recovery force, and the individual. Army forces are typically equipped and prepared to survive for up to 96 hours of isolation subject to recovery force capabilities, threats, and environmental conditions. ATP 3-50.20, chapter 2 provides an in-depth discussion on duration of isolation.

NAVIGATION

1-8. Navigation requirements for isolated personnel are met through planning and preparation efforts that focus on a combination of Global Positioning System (GPS), map and compass, and field-expedient techniques in a specific operational environment for the purpose of surviving, avoiding capture, and facilitating recovery. Navigational limitations and opportunities relating to terrain and illumination are identified and analyzed to subsequently organize, train and equip personnel to conduct precise movements to desired locations such as recovery areas, recovery sites and rally points to enable link-up with recovery forces. Assess navigation needs during planning and preparation activities and integrated into organization, training, and equipping considerations by the commander and staff, unit recovery force, and the individual.

PSYCHOLOGY OF SURVIVAL

1-9. It takes much more than the knowledge and skill to build shelters, gather food, build fires, and move without the aid of standard navigational devices, to survive isolation and possible detention or captivity. The key ingredient in any survival situation is the mental attitude of the isolated person involved. Having survival skills is important; having the will to survive is essential. Without a desire to survive, acquired skills serve little purpose and invaluable knowledge goes to waste.

1-10. During isolation, isolated personnel will experience numerous stressful moments that will shape them for the remainder of their lives. Stress is not a disease that you cure and eliminate. Instead, we all experience a condition. Stress is our reaction to pressure. It is the name given to the experience we have as we physically, mentally, emotionally, and spiritually respond to life's tensions.

1-11. Stress compels one to act or prepare to respond to any given situation. Humans need stress because it can have many positive benefits. Stress-inducing situations provide humans with challenges and give them chances to learn about their values and strengths. Stress can show them their ability to handle pressure without breaking. It tests their adaptability and flexibility, and can stimulate and motivate them to do their best. The goal is to have stress, but not in excess. Too much stress leads to distress. Distress causes an uncomfortable tension that a person tries to escape or avoid. Here are a few of the common signs of distress that isolated personnel will encounter during isolation:

- Difficulty making decisions.
- Forgetfulness.
- Low energy level.
- Constant worrying.
- Propensity for mistakes.
- Carelessness.
- Trouble getting along with others.
- Withdrawing from others.
- Angry outbursts.
- Hiding from responsibilities.
- Thoughts about death or suicide.

1-12. Stress can encourage or discourage, move us along or stop us dead in our tracks, and make life meaningful or seemingly meaningless. It can inspire us to operate successfully and perform at maximum efficiency. It can also cause us to panic and forget all common sense and training. One key to survival is the isolated person's ability to manage the inevitable stresses that they will encounter.

SURVIVAL STRESSORS

1-13. Once the body recognizes the presence of a stressor; it protects itself and/or address the stressor. Often, stressful events occur simultaneously. In response to a stressor, the body prepares to either "fight, flight or freeze." This causes the following coping reactions that cannot be maintained for long periods:

- The body releases stored fuels (sugar and fats) to provide quick energy.

- Breathing rate increases to supply more oxygen to the blood.
- Muscle tension increases to prepare for action.
- Blood clotting mechanisms activate to reduce bleeding from cuts.
- Senses become more acute (hearing becomes more sensitive, pupils dilate, smell becomes sharper) so that you are more aware of your surroundings.
- Heart rate and blood pressure rise to provide more blood to the muscles.

1-14. The cumulative effect of stress and stressors can add up if not managed properly. As the body's resistance to stress wears down and the sources of stress continue (or increase), a state of exhaustion arrives. At this point, the ability to resist stress or use it in a positive way gives out and signs of exhaustion appear. Anticipating stressors and developing strategies to cope with them are two ingredients in the effective management of stress. Therefore, it is essential that you are aware of the types of stressors that you could encounter if isolated. Injury, illness, disability, and death are real possibilities that one may have to face.

PREPARING YOURSELF

1-15. The isolated person's mission is to stay alive. The assortment of thoughts and emotions isolated persons will experience in a survival situation can work for them, or against them. Fear, anxiety, anger, frustration, guilt, depression, and loneliness are all possible reactions to the many stressors common to survival. These reactions, when managed in a healthy way, help to increase the likelihood of surviving. They prompt people to pay more attention in training, to fight back when scared, to take actions that ensure sustenance and security, to keep faith with their fellow team members, and to strive against large odds. When isolated personnel cannot control these reactions in a healthy way, they can bring them to a standstill. Instead of rallying their internal resources they listen to their internal fears, which causes the isolated person to experience psychological defeat long before they physically succumb. Remember, survival is natural to everyone; being unexpectedly thrust into the life-or-death struggle of survival is not.

BE REALISTIC

1-16. Isolated personnel should not be afraid to make an honest appraisal of the situation. They should see circumstances as they are, not as they might want them to be, while keeping hopes and expectations within their estimate of the situation. When going into a survival setting with unrealistic expectations, one may be laying the groundwork for bitter disappointment. Follow the adage, "Hope for the best, prepare for the worst." It is much easier to adjust to pleasant surprises about unexpected good fortunes than to be confronted with unanticipated harsh circumstances.

ADOPT A POSITIVE ATTITUDE

1-17. Isolated personnel should learn to see the potential good in everything. While looking for the good not only boosts morale, it also is excellent for exercising their imagination and creativity. While trying to take ordinary objects and figuring out different uses for them in a survival situation, everything is a potential tool.

TRAIN AND PREPARE

1-18. Through military training and life experiences, begin today to prepare to cope with the rigors of survival. When you are capable of demonstrating your skills in training today you will have the confidence in your ability to call upon them if the need should arise. The goal of preparation is to build confidence in your ability to function despite your situation and fears. Failure to prepare yourself physically and psychologically to cope with isolation leads to reactions such as depression, carelessness, inattention, loss of confidence, and poor decision-making. Remember that your life and the lives of others who depend on you are at stake.

SURVIVAL PATTERN

1-19. The survival pattern is a tool that enables personnel to mitigate the effects of being an in an isolation situation. The survival pattern includes the manipulation of applicable sustenance, protection, navigation and survival medicine skills placed in order of priority. For example, in a cold environment, fire would be the first priority to get warm; followed by a shelter to protect from the cold, wind, and rain or snow; access to or the ability to procure potable water; traps or snares to get food; and survival medicine to maintain health. In all isolation situations, survival medicine has top priority.

1-20. In Figure 1-1 on page 26, the word SURVIVAL is a useful memory tool to aid isolated personnel in making sound decisions on what has to happen to meet their needs and appropriate actions.

Figure 1-1. Survival Acronym

S **Size up the Situation**
(Surroundings, Physical Condition, Equipment)

U **Use all Your Senses,**
Undue Haste Makes Waste

R **Remember Where You Are**

V **Vanquish Fear and Panic**

I **Improvise**

V **Value Living**

A **Act Like the Natives**

L **Live by Your Wits, But for Now, Learn Basic Skills**

1-21. Building and maintaining situational understanding is essential for isolated personnel to understand their isolation situation, develop effective plans, make quality decisions and execute their ISG[1] or EPA.[2]

1-22. The typical isolation scenario includes the occurrence of an isolating event and the typical movement of personnel to their designated rally point. Once at the rally point, they assess their situation in relation to the mission and the isolation criteria specified

---

1. Isolated soldier guidance.
2. Evasion plan of action

in their ISG/EPA. The isolation criteria prescribes the set of circumstances under which a Soldier/isolated person may execute their ISG/EPA. The assessment of the isolation criteria includes the operational political, military, economic, social, information, infrastructure, physical environment, and time (PMESII-PT), as well as mission variables mission, enemy, terrain and weather, troops and support available, time available, and civil considerations (METT-TC) and comprehension of the factors of isolation, individual capability (i.e. physical and psychological condition, proficiency of training, equipment), effects posed by an operational environment and the potential duration of the isolation situation. If the isolated person determines that they have met the isolation criteria stated in their ISG/EPA, they will report their isolation and begin appropriate movements and techniques detailed in their ISG/EPA.

1-23. Isolated personnel continue to adjust their plan to meet survival medicine, protection, sustenance, and navigation needs in order of priority and update their situational understanding as key tactical factors change while making appropriate adjustment decisions.

CHAPTER 2

SURVIVAL MEDICINE

Isolated persons require knowledge of many different survival skills to return to friendly forces. This chapter will discuss the psychology of survival, treatment for shock, and survival medicine practices.

SURVIVAL MEDICINE VERSUS TRADITIONAL MEDICINE

2-1. Excluding the enemy, medical related problems arising from combat and isolation pose the greatest threat to isolated personnel. You must understand the vast differences and transition of medical care that take place following combat, isolating events, and implementation of your ISG/EPA. Isolated personnel must be able to perform fundamental survival medicine techniques throughout the duration of their isolation including during evasion, detention or captivity, and recovery. During combat operations, injured Soldiers follow Tactical Combat Casualty Care (TCCC) guidelines and protocols focused on the care of casualties in a combat or tactical environment at the point-of-injury. The TCCC program of instruction prepares Soldiers to provide self-aid or buddy-aid in the absence of a medical provider. Four transitions of medical care occur on the battlefield that includes isolation producing events.

- Perform Care under Fire (CUF) at the point-of-injury on the battlefield. The Soldier is typically equipped with the Improved First Aid Kit (IFAK), Combat Pill Kit, and Eye Shield. The Soldier returns fire and takes cover while focused on the treatment of massive hemorrhage.

- Perform Tactical Field Care when no longer under hostile fire. The Soldier is equipped with the Individual First Aid Kit (IFAK), Combat Pill Kit, and Eye Shield. The Soldier is subject to a reduced level of hazard from hostile fire, more time is available to provide care based on the tactical situation, a security perimeter is established and the casualty evaluated for altered mental status. Follow the MARCH algorithm as a guide to the sequence of treatment priorities in caring for combat casualties:

- **M**assive hemorrhage — control life threatening bleeding.
- **A**irway — establish and maintain a patent airway.[1]
- **R**espiration — decompress suspected tension pneumothorax, seal open chest wounds, and support ventilation/oxygenation as required.
- **C**irculation — establish intravenous (IV) access and administer fluids as required to treat shock.
- **H**ead injury/Hypothermia — prevent/treat hypotension and hypoxia to prevent worsening of traumatic brain injury and prevent/treat hypothermia.

• Perform medical care at the point-of-injury throughout isolation. Typically, the Soldier's IFAK and supporting medical supplies are already expended; therefore, Survival medicine takes over at this point. Survival medicine recognizes the fundamentals of the MARCH algorithm and provides a flexible approach that prioritizes actions relevant to physical, psychological, and environmental considerations that are continually assessed and prioritized for performance by isolated personnel.

2-2. Survival medicine requires isolated personnel to—

• Understand the fundamentals of trauma first aid (MARCH) and survival medicine.
• Assess and prioritize applicable fundamentals from trauma first aid and survival medicine in a resource-constrained environment.
• Execute those fundamentals under sub-optimal healing conditions.
• Continually assess your mental state and enable appropriate adjustments.
• Understand that medical personnel and facilities are rarely available.

2-3. Survival medicine employs four techniques to facilitate isolated personnel's performance of survival proficiencies. These techniques include—

• **Prevention**. The common sense act of proactive prevention or hindrance of an action that could lead to a requirement to perform trauma first aid or survival medicine. For example, during evasion movement, the Soldier looks for alternate crossing points across a river instead of swimming across to decrease the chance of getting hypothermia and having to dry their clothing.

---

1. A "patent" airway is one in which no obstructions exist between the subject's lungs and the outside air.

- **Recognition**. The act of actively recognizing—
  - Symptoms (urine color and relation to dehydration).
  - Capabilities (identify local medicinal plants such as cattails, willow during movement).
  - Identity (venomous snakes, insects, etc.)
- **Mitigation**. The actions taken to immediately reduce the severity and or pain associated with a survival related injury or illness (application of a tourniquet to stop severe, uncontrolled bleeding that could cause loss of life)
- **Treatment**. The actions taken to manage and care for the isolated person with a disease or disorder and restore their health. The ability of the isolated person or personnel to plan, prepare, execute and assess appropriate actions to meet their survival needs significantly increases their overall chances of survival, increases morale, and aids in their ability to perform the survival proficiencies and eventual return to friendly forces.

2-6.[1] Become familiar with the survival medicine capabilities available in the Survival Kit. Prior to operations, personnel assess the survival kit to identify its survival medicine components. It is recommended that the Soldier rehearse appropriate survival medicine techniques using similar materials as those found in the actual survival kit. The survival medicine capabilities that should be considered for inclusion in the survival kit includes these items:

- Water purification tablets.
- Small tube of antibiotic ointment.
- 1 oz. bottle of 2% tincture of iodine.
- Small tube of Crazy Glue/Super Glue.
- Small tube/bottle of Betadine (povidone-iodine).
- Emergency blanket.
- Floss card (dental floss).
- Small roll of duct tape with peel-away backing.
- Heavy duty canvas sewing needle.
- Assorted dressings.
- Assorted bandages.[2]
- Combat Gauze.
- SOFT-T[3] or combat application tourniquet (CAT)[4] or Israeli bandage.[5]

---

1. Paragraphs 2-4 and 2-5 have been removed due to repetition (see Paragraphs 2-160~2-163).
2. See Paragraph 2-24 for information on field-expedient dressings and bandages.
3. SOF Tactical Tourniquet.

- Small package of prescription medications, contacts.
- Gear that accomplishes more than one task (for example, use a large bandana as a compress, sling, bandage, and eye patch).

LIFESAVING STEPS

2-7. Once the Soldier has implemented their ISG/EPA, they become their own doctor, emergency medical technician, and surgeon; in an operational environment where they are being hunted, and without any of the medical capabilities that are typically needed to treat the recognized condition/symptoms. The isolated person's health is of primary importance; taking unnecessary risks, which could lead to injury, are prevented or mitigated. Once isolated, the fundamentals of the MARCH algorithm (see page 29) are applied through a flexible process that prioritizes actions relevant to physical, psychological, and environmental considerations that are continually assessed and prioritized for action by the isolated person.

2-8. The following list of actions enable assessment and lifesaving treatment of isolated persons. The list is a guide that makes use of the MARCH algorithm (see page 29). If the isolated person does not exhibit "massive hemorrhage", the priority for treatment follows the Airway, Respiration (breathing), Circulation, and Head/Hypothermia protocol.

- In a survival situation, if you discover another casualty you should determine if the casualty is alive or dead. If there are no signs of life — no pulse, no breathing — you should NOT attempt to perform lifesaving steps. You should note the Soldier's name, rank, and location. You should perform field recovery of equipment from the casualty that will aid your survival effort.
- If the casualty is alive you should provide care to the casualty using prioritized actions from Battle Drill 3 "Perform Tactical Combat Casualty Care" referenced in Appendix A, STP 21-1-SMCT, 28 Sept. 2017.

Note: Steps 3 through 13 (from the Soldier's Manual of Common Tasks) are performed as self-aid for the isolated person and buddy-aid for the casualty/other isolated personnel.

---

4. See Paragraphs 2-12 and 2-13 for information on improvising a tourniquet when a CAT, SOFT-T, or similar is unavailable.
5. Also known as the Emergency Bandage, the Israeli Bandage incorporates a built-in pressure applicator to stop bleeding from hemorrhagic wounds.

TREATING BLEEDING

2-9. Identify external, life-threatening bleeding. Bleeding is life threatening if any one of the following signs/symptoms are observed:

- There is a traumatic amputation of an arm or leg.
- There is pulsing or steady bleeding from the wound.
- Blood is pooling on the ground.
- The overlying clothes are soaked with blood.
- Bandages or makeshift bandages used to cover the wound are ineffective and steadily becoming soaked with blood.
- There was prior bleeding, and the casualty is now in shock (unconscious, confused, pale).

2-10. Control external bleeding. External bleeding falls into the following classifications:

- **Arterial**. Blood vessels called arteries carry oxygenated blood away from the heart and throughout the body. Arterial bleeding is the most serious type of bleeding. If not controlled promptly, it can be fatal. With this type of bleeding, the blood is typically bright red to yellowish in color and exits the wound in distinct spurts or pulses that correspond to the rhythm of the heartbeat rather than in a steady flow. Because the blood in the arteries is under high pressure, an individual can lose a large volume of blood in a short period.
- **Venous**. Venous blood is blood that is returning to the heart through blood vessels called veins. A steady flow of dark red, maroon, or blackish in color blood characterizes bleeding from a vein due to the lack of oxygen it transports. Venous bleeding is still of concern. While the blood loss may not be arterial, it can still be quite substantial, and can occur with surprising speed if not treated. It can usually be controlled more easily than arterial bleeding.
- **Capillary**. The capillaries are the extremely small vessels that connect the arteries with the veins. Capillary bleeding most commonly occurs in minor cuts and scrapes and generally oozes in small amounts as opposed to squirting (arterial) or flowing (venous). This type of bleeding is not difficult to control.

2-11. If the casualty has severe, life-threatening bleeding from an extremity or has an amputation of an extremity, administer life-saving hemorrhage control by applying a combat application tourniquet (CAT) from the casualty's IFAK before moving the casualty. Personnel with life-threatening bleeding can bleed to death from a complete femoral artery and vein disruption within as little as three minutes. Isolated persons must control life-threatening bleeding immediately because replacement fluids are not available.

2-12. If a CAT is unavailable, apply an improvised tourniquet made from a rod (made from a jack handle, stick, scabbard, cleaning rod, pipe, dowel), a band of material at least 1-1/2 inches wide (made from a cravat, bandana, towel, ace bandage, shirt, nylon webbing, rifle sling.

Note: Belts or zip ties should only be used as a last resort, and a securing mechanism as a constricting or compressing device to control arterial and venous blood flow to a damaged extremity for a short period of time.

2-13. The wide band of material is made into a loop that fits over the damaged limb, 2-3 inches above the site of arterial bleeding, and tied tightly with an overhand knot. Next, lay the rod across the overhand knot. The running ends of the loop are then used to tie another overhand knot on top of the rod forming a square knot with the rod through the center of the knot. The rod is then twisted, applying pressure circumferentially around the limb tight enough to stop the arterial bleeding. Do not tighten the tourniquet more than necessary to stop the bleeding. In the case of amputation, dark oozing blood may continue for a short time. This is the blood trapped in the area between the wound and tourniquet. Fasten the tourniquet to the limb by looping the free ends of the tourniquet over the ends of the stick. Then bring the ends around the limb to prevent the stick from loosening. Tie them together on the side of the limb.

2-14. If bleeding remains, place a second tourniquet side by side to the first tourniquet. A tourniquet can be left in place up to two hours without damage to vessels, nerves, muscle or loss of limb. If isolated, the victim or buddy should release the pressure from the tourniquet after two hours, and then retighten if blood loss continues.

2-15. Ideally, the tourniquet will stop or considerably slow down the flow of arterial blood from the wound. As an open wound, the risk of infection is great. Before applying a pressure bandage, rinse the wound with sterile saline or clean water.

Note: In isolation or captivity; alcohol, vinegar, natural honey, hydrogen peroxide, and bleach are also highly effective antiseptics. Once gauze or bandage is applied it must not be removed.

TREATING HYPOXIA

2-16. Hypoxia is the result of insufficient oxygen in the blood. It is a potentially deadly condition and one of the leading causes of cardiac arrest. Cardiac arrest is linked to an absence of circulation in the body, for any one of a number of reasons. For this reason, maintaining circulation is vital to moving oxygen to the tissues and carbon dioxide out of the body. Open an airway and maintain it by using the following steps:

- Check to see if the casualty has a partial or complete airway obstruction. If they can cough or speak, allow them to clear the obstruction naturally. Stand by, reassure the casualty, and be ready to clear their airway and perform mouth-to-mouth resuscitation should they become unconscious. If their airway is completely obstructed, administer abdominal thrusts until the obstruction is cleared. Any one of the following can cause airway obstruction, resulting in stopped breathing:
  - Foreign matter in mouth of throat that obstructs the opening to the trachea.
  - Face or neck injuries.
  - Inflammation and swelling of mouth and throat caused by inhaling smoke, flames, and irritating vapors or by an allergic reaction.
  - "Kink" in the throat (caused by the neck bent forward so that the chin rests upon the chest).
  - Tongue blocking passage of air to the lungs upon unconsciousness. When an individual is unconscious, the muscles of the lower jaw and tongue relax as the neck drops forward, causing the lower jaw to sag and the tongue to drop back and block the passage of air.
- Using a finger, quickly sweep the casualty's mouth clear of any foreign objects, broken teeth, dentures, sand, etc.
- Using the jaw thrust method (Figure 2-1), grasp the angles of the casualty's lower jaw and lift with both hands, one on each side, moving the jaw forward. For stability, rest your elbows on the surface on which the casualty is lying. If their lips are closed, gently open the lower lip with your thumb.

Figure 2-1. Jaw Thrust Method

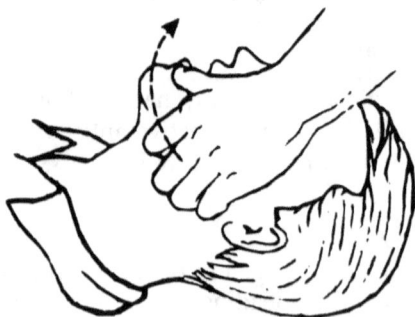

- Grasp the angles of the lower jaw and lift with both hands, one on each side, moving the jaw forward.

- If victim's lips are closed, open the lower lip with your thumb.

- With the casualty's airway open, pinch their nose closed with your thumb and forefinger and blow two complete breaths into their lungs. Allow the lungs to deflate after the second inflation and perform the following:

- Look for the chest to rise and fall.
- Listen for escaping air during exhalation.
- Feel for flow of air on your cheek.

2-17. If the forced breaths do not stimulate spontaneous breathing, maintain the casualty's breathing by performing mouth-to-mouth resuscitation.

2-18. There is a danger of the victim vomiting during mouth-to-mouth resuscitation. Check the victim's mouth periodically for vomit and clear as needed.

2-19. If the casualty is unconscious, if respiratory rate is less than 2 breaths in 15 seconds, and/or if the casualty is making snoring or gurgling sounds, insert a nasopharyngeal airway (NPA)[1] from the casualty's IFAK. Remember these things when inserting the NPA:

- Keep the casualty in a face-up position.
- Lubricate the tube of the NPA with water.
- Push the tip of the casualty's nose upward gently.
- Position the tube of the NPA so that the bevel (pointed end) of the NPA faces toward the septum (the partition inside the nose that separates the nostrils).
- Insert the NPA into the nostril and advance it until the flange rests against the nostril.

Note: Cardiopulmonary resuscitation (CPR) may be necessary after cleaning the airway, but only after major bleeding is under control.

2-20. Continue to check for bleeding by performing a blood sweep.[2] Control external bleeding by the application of direct pressure, indirect pressure, elevation, or digital ligation (finger pressure).

DIRECT PRESSURE

2-21. The most effective way to control external bleeding is by applying pressure directly over the wound. This pressure must not only be firm enough to stop the bleeding, but it must also be maintained long enough to "seal off" the damaged surface.

2-22. If bleeding continues after having applied direct pressure for 30 minutes, apply a pressure dressing. This dressing consists of a thick dressing of gauze or other suitable material applied directly over the wound and held in place with a tightly wrapped bandage. It

---

1. The nasopharyngeal airway (NPA), also known as a nasal trumpet or nose hose, is a tube or catheter with one flared end, flange or washer. The purpose of the flared end is to secure the NPA against the subject's nose, preventing it from being sucked or falling entirely into the nasal cavity.

should be tighter than an ordinary compression bandage but not so tight that it impairs circulation to the rest of the limb.

2-23. Once you apply the dressing, do not remove it, even when the dressing becomes blood-soaked. Leave the pressure dressing in place for 1 or 2 days, after which you can remove it and replace it with a smaller dressing. In a long-term survival environment, make fresh, daily dressing changes and inspect for signs of infection. Figure 2-2 on page 37 shows applications for pressure dressings.

---

2. A blood sweep is an inspection of the subject's body with the purpose of rapidly identifying life-threatening massive bleeding by visual and physical inspection of the subject, particularly in situations where such bleeding may not be immediately obvious (due to thick clothing, low-light conditions, etc.) To perform a blood sweep, palpate (gently press with the fingertips) the subject's body with your (clean) hands, starting at the head and neck. Feel for signs of bleeding (wetness) and every few seconds inspect your hands for blood. Inspect the subject's body visually during this palpation. Continue this process on the rest of the subject's body from head to toe, including the extremities, not neglecting the armpits and groin, checking for signs of bleeding every few inches. If the subject is lying down, slide your hands under the subject's body to check for bleeding there.

Figure 2-2. Application of a Pressure Dressing

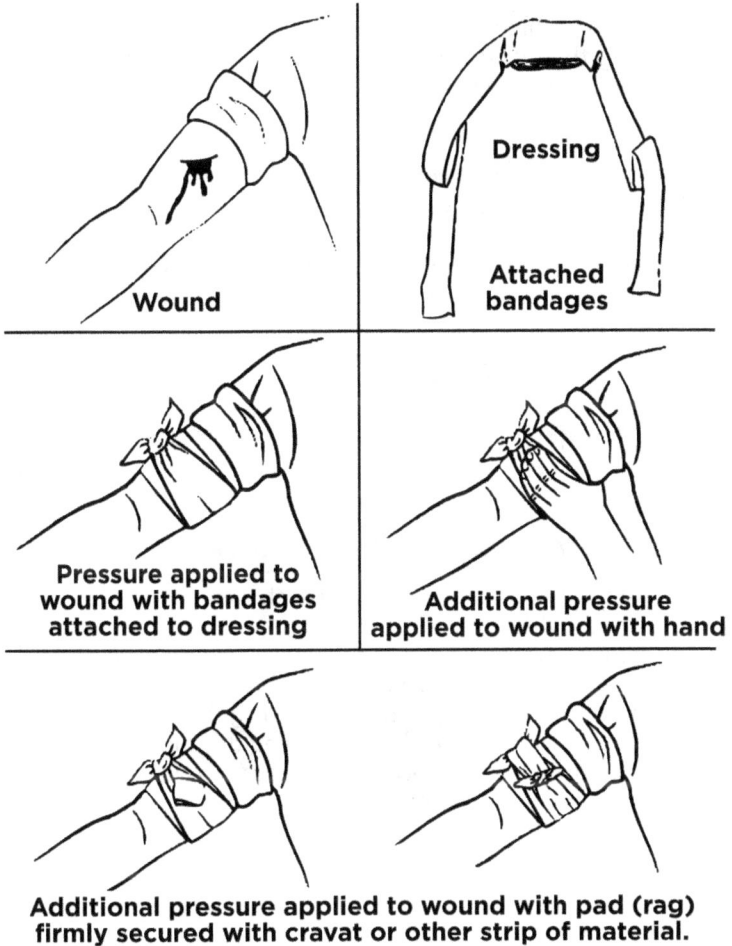

Additional pressure applied to wound with pad (rag)
firmly secured with cravat or other strip of material.

2-24. You can make field-expedient dressings if a medical kit is unavailable. The purpose of a dressing is to control bleeding, absorb wound secretions, and to prevent bacteria from entering the wound. Materials that make functional field-expedient dressings include cloth from a shirt, undergarments, socks, bandanas, handkerchiefs, thin towels, bedding and feminine care absorbent pads. Cut these materials to proper size to cover the wound and sterilized before use. To sterilize, you can steam the material for five minutes or boil the material in water for ten full minutes at a rolling boil. If needed, you can clean and sanitize used bandages by boiling them, then reusing them (if no other option exists).

ELEVATION

2-25. Raising an injured extremity as high as possible above the heart's level slows blood loss by aiding the return of blood to the heart and lowering the blood pressure at the wound. However, elevation alone will not completely control bleeding; apply direct pressure over the wound.

PRESSURE POINTS

2-26. A pressure point is a location where the main artery to the wound lies near the surface of the skin or where the artery passes directly over a bony prominence. Personnel can use digital pressure on a pressure point to slow arterial bleeding until a pressure dressing is applied. Pressure point control is not as effective for controlling bleeding as direct pressure exerted on the wound. It is rare when a single major compressible artery supplies a damaged vessel.

WARNING: Use caution when applying pressure to the neck. Too much pressure for too long may cause unconsciousness or death. Never place a tourniquet around the neck.

2-27. If you cannot remember the exact location of the pressure points, you should follow this rule: Apply pressure at the end of the joint just above the injured area. On hands, feet, and head, this will be the wrist, ankle, and neck, respectively. Maintain pressure points by placing a round stick in the joint, bending the joint over the stick, and then keeping it tightly bent by lashing. Using this method to maintain pressure frees the hands to work in other areas. Figure 2-3 on page 39 shows key body pressure points.

Figure 2-3. Key Body Pressure Points

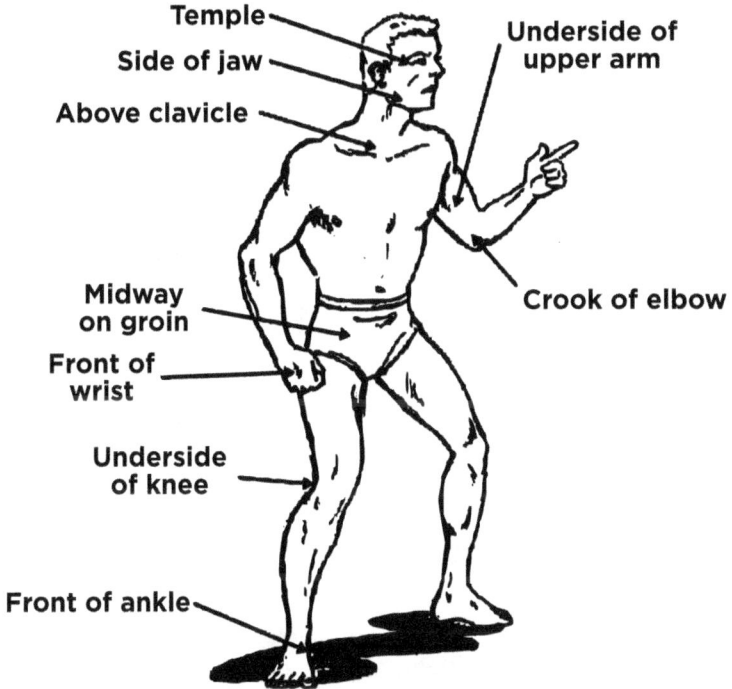

Temple

Side of jaw

Above clavicle

Underside of
upper arm

Midway
on groin

Front of
wrist

Crook of elbow

Underside
of knee

Front of ankle

2-28. If the bleeding is on the head, above the ears, press the point just in front of the ear, in a direct line to the corner of the eyes.

2-29. If the lower part of the face is bleeding, press the point on the jaw bone halfway between the chin and the end of the jaw.

2-30. If bleeding is from the neck, press the point on the carotid artery, located between the Adam's Apple and neck muscles. Stopping bleeding from here is a matter of life and death.

2-31. If the bleeding is high on the arm, press the point just above the middle of the collar bone. If the bleeding is low on the arm, press the point in the fold opposite the elbow, on the inside of the arm.

2-32. There are two pressure points on the wrists. The first one is more common, where we normally feel our pulse. The other one is just alongside, down from the little finger.

2-33. If the bleeding is from the groin or thighs, find and press the femoral artery. It is located along the bikini line, half way between the hip and the groin. A lot of pressure is required to control the bleeding here, maybe even both your hands, due to the amount of blood that flows through this artery to supply oxygen to the legs.

2-34. The pressure point on the popliteal artery lies behind the knee. Press this to stop bleeding from the lower leg below the knee.

DIGITAL LIGATION

2-35. Slow down major bleeding by applying pressure with a finger or two on the bleeding end of the vein or artery. Maintain the pressure until the bleeding stops or slows down enough to apply a pressure bandage, and elevation.

ASSESS FOR BREATHING AND CHEST INJURIES

2-36. Fractured ribs are common, painful, and disabling. The isolated person will not have access to pain medications and must understand that the pain associated with rib injuries can lead to reduced movement and cough suppression which can contribute to formation of secondary chest infection. To treat fractured ribs —

- Protect the injured rib by supporting the arm on the injured side with a sling-and-swathe.
- Encourage the person to take deep breaths regularly, even if it hurts, to keep the lungs clear.
- Watch the person for increasing trouble breathing.

CAUTION: Do not wrap a band snugly around the person's chest.

2-37. Flail chest. It is most commonly a result of serious blunt trauma (falling from a height, vehicle or aircraft wreck or other accident). Ribs are typically broken away from the sternum, or when two, three or more adjoining ribs are broken in two or more places. The condition will make breathing difficult and indicates possible internal bleeding. Detect "flail chest" by observing a section of ribs moving in and out opposite to the rest of the ribs during breathing due to air pressures. You can try applying a bulky dressing and wrapping it to immobilize.

2-38. Pneumothorax. This is a common injury in isolation where personnel fall or suffer chest trauma associated with blast injuries, blunt trauma and penetrating trauma (sucking chest wound). Sucking chest wounds are recognized by the sucking noise and appearance of foam or bubbles in the wound. This condition requires application of an occlusive dressing[1] to the entry and exit wounds immediately before serious respiratory and circulatory complications occur.

2-39. Ideally, the patient should attempt to exhale while holding the mouth and nose closed (the Valsalva maneuver, as used to clear or equalize the ears after a pressure change such as when diving or changing altitude) as the wound is closed. This inflates the lungs and reduces the air trapped in the pleural cavity. Frequently, a taped, airtight dressing is all that is needed, but sometimes it is necessary to suture the wound to make sure the wound is closed.

2-40. Begin rescue breathing as necessary to restore breathing and/or pulse (Cardiopulmonary Resuscitation (CPR)).

- Place the casualty on a firm, flat surface. Give 30 chest compressions by compressing the casualty's chest at least 2 inches deep.
- Push hard, push fast in the middle of the chest at a rate of at least 100 compressions per minute.
- Give 2 rescue breaths by tilting the head back and lift the chin up, then pinch the nose shut then make a complete seal over the person's mouth.
- Blow in for about 1 second to make the chest clearly rise. Give rescue breaths, one after the other. If chest does not rise with the initial rescue breath, retilt the head before giving the second breath.
  - If the second breath does not make the chest rise, the person may be choking. After each subsequent set of chest compressions and before attempting breaths, look for an object and, if seen, remove it. Continue CPR.

TREATING BURNS

2-41. The following treatments are extreme measures and are only meant to be applied with extreme caution under real-world experiences. Burns sustained during military operations constitute a relatively small, but very real percentage (5%) of combat-related injuries. Even burns to a small surface area can be incapacitating for the casualty and strain the resources of deployed military medical units. It is crucial to remember that burns may represent only one of the casualty's traumatic injuries, particularly when an explosion is the mechanism of injury. Optimal treatment includes management of secondary conditions (such as shock and hypothermia) related to the burn and associated traumatic injuries. Resuscitation of the burn casualty is generally the most challenging aspect of care during the first 48 hours following injury, and optimal care requires a concerted effort on the part of all providers involved during the evacuation and treatment process.

---

1. An air- and water-tight seal. A key objective of treating a sucking chest wound is to prevent air from entering the chest cavity. Some manufactured chest seals (such as the Asherman chest seal) possess a one-way valve, allowing air and fluids to exit the chest cavity but not to re-enter. If a manufactured chest seal is unavailable an occlusive dressing can be improvised using any (clean) plastic material to hand, such as the packaging of other medical supplies, using the duct tape (or similar) from your survival kit to seal the dressing.

2-42. Dress all non-life threatening injuries and any bleeding that has not been addressed earlier with appropriate dressings. Also—

- Check the casualty for burns. Burns are painful and limit capability and can increase the susceptibility of shock and infection and lead to a loss of body heat, fluids, and minerals.
- If an isolated person catches on fire, the immediate action is to get away from the flame and smother the fire—do not run—Stop, Drop, and Roll.
- Suspect possible airway complications with burns to the face and/or neck; soot in the mouth and or nose; singed facial hair; or a dry cough.
- If there is a chemical causing the burn, remove the chemical from the skin by flushing it with copious amounts of water.
- In case of electrical burns, remove the victim from the contact with electricity. Electrical burns typically affect cardiac or respiratory systems. Always care for cardiac and respiratory problems before caring for burns.

2-43. Once the fire is out, assess the damage and begin to treat the burns. Initiate MARCH (see page 29) and protect the casualty from shock and hypothermia. Cool the burning skin with cold water for 20 minutes if possible. For burns caused by white phosphorous, pick out the white phosphorous with tweezers; do not douse with water.

2-44. First-degree burns involve only the outer layer of the skin known as the epidermis. First-degree burns do not blister; they become very red and are quite painful. After about two to three days, the pain should subside and peeling of the skin will begin. Soaking in cool water helps with the pain.

2-45. Second-degree burns are known as superficial partial-thickness burns. These types of burns affect the upper layers of the dermis and have a tendency to swell and blister. They are more painful than first-degree burns. If you suspect a second-degree burn:

- Remove jewelry or tight clothing from the burned area before the skin begins to swell.
- Take a pain reliever if available. Flush the skin with cool running water or apply moist cloths until pain lessens.
- Do not use ice or ice water; this can cause more damage to the skin.
- Use an antiseptic spray to prevent infection or use aloe cream or the aloe plant to soothe the skin.
- Do not put greasy ointments, butter, or similar substances on a burn. Grease could lower the rate of heat release from the skin, worsening the burn's effects.
- Cover the burn with a clean, dry, non-adhering no-fluff bandage such as a gauze pad.

2-46. Third-degree burns or full-thickness burns involve all layers of the dermis. This type of burn causes a large amount of tissue damage and is extremely painful (or painless if nerve damage has occurred). Skin will be leathery and dry and can appear black, white, or brown in color. Do not remove any embedded charred clothing or material. However, remove constrictive jewelry and unburned clothing from the area if they are not stuck on the burn. If you suspect a third-degree burn—

- Apply cool, wet compresses to the burned area for a very brief time to help reduce the body temperature. Do not use ice or immerse the affected area in cold water.
- Cover the burned area with cool, moist, sterile bandages.

2-47. Be prepared to treat injured personnel for shock. If possible, elevate the burn above the heart to assist in reducing swelling and the likelihood or severity of shock. Staying hydrated will help control the loss of body fluids. Isolated persons should replace salt by consuming the cooked eyes and blood of animals or adding ¼ teaspoon of regular salt per quart of water.

2-48. Prevent/treat hypotension and hypoxia to prevent worsening of traumatic brain injury (TBI) and prevent/treat hypothermia.

TREATING HEAD INJURY

2-49. During Tactical Field Care, Soldiers assess their condition for an altered mental status. Soldiers with an altered mental status may use their weapons or radios inappropriately and may not be able to accurately assess their condition in relation to isolation criteria and implementation of their ISG/EPA. Injuries to the head pose additional problems related to brain damage and may interference with breathing and eating. Bleeding is more profuse in the face and head area, but infections are less likely. This makes it somewhat safer to close such wounds earlier to maintain function.

2-50. Most important in treating a head injury is maintaining proper airway control. The lack of oxygen to an injured brain can have very detrimental effects. An emergency airway puncture may be necessary if breathing becomes difficult due to obstruction of the upper airways. In the event of unconsciousness, keep the patient still and under close observation. Even in the face of mild or impending shock, keep the head level or even slightly elevated if there is reason to expect brain damage. Do not give fluids or morphine to unconscious persons.

2-51. Traumatic Brain Injury (TBI) is a common injury in isolation where personnel fall or suffer head trauma associated with vehicle and aircraft crashes, blast injury, and blunt trauma. Hypotension (low blood pressure) and hypoxia (insufficient oxygen in the blood) produce dizziness and fainting.

2-52. For a bleeding scalp, apply several dressings with your gloved hand. Press gently because the skull may be fractured. On examination, if you feel a depression, spongy area or bone fragments, DO NOT put direct pressure on the wound. Control bleeding with a gauze dressing and general, distributed pressure.

2-53. The symptoms associated with mild brain injuries typically include short term memory loss, blurred vision, nausea, dizziness, extreme tiredness and possible neck pain or tenderness. Care for mild brain injury includes—

- Apply pressure from a gauze dressing on the bleeding scalp.
- Apply a cold pack, ice, wet bandana, to reduce swelling and pain associated with a bump.
- Monitor the person for 24 hours.
- Awaken the person every 2 hours to check for signs and symptoms of serious brain damage.

2-54. The symptoms associated with serious brain injury include—

- Prolonged unconsciousness with no response to aggressive stimulation, such as shouting or tapping the shoulder.
- Possible skull fracture. Symptoms include—
  - A depression (pressed-in area) in the skull. (DO NOT push on the area.)
  - A fracture that is visible where the scalp has been torn.
  - Bruising around both eyes (raccoon eyes) or behind both ears
  - Clear fluid and/or blood dripping from the nose or ears.
- Inability to sense touch or move extremities.
- Eyes that do not respond to light appropriately or equally. (Check pupil response, one eye at a time, by shading the eyes with a hand and then exposing the pupils to light.)
- Mental status deterioration (from disorientation, to irritability, to combativeness, to coma).
- Personality changes.
- Loss of coordination, balance and/or speech.
- Extremely bad headaches.
- Vision problems.
- Seizures.
- Nausea and vomiting that does not go away.
- Relapsing into unconsciousness.
- Heart rate that slows down to less than 40 beats per minute (BPM), then speeds up.
- Erratic (irregular) respiratory rate.
- Unequally sized pupils.

2-55. Care for serious brain injury includes—

- Assessing the person for spinal injury.
- Keeping the person calm and reassured.

- Using two personnel to help the casualty move if they are able to walk.
- Being prepared to treat the person for reduced breathing.

2-56. In a situation where a group of personnel are isolated, consideration should be given to removing all weapons and radios away from any casualty manifesting brain injury who is not alert and fully oriented to the tactical situation. The group must exercise the tenets of mission command and ensure that appropriate decisions about ISG/EPA, captivity, and detention are also implemented.

2-57. The individual isolated person with a brain injury should consider immediate movement to and occupation of a hide site to rest and relieve their condition.

2-58. Administer pain medications and antibiotics from the combat pill pack if available.

TREATING SHOCK

2-59. Monitor the patient for shock and treat as appropriate. Shock is a condition of the body that describes the physiologic condition where oxygen delivery to the tissues of the body is not enough to meet the metabolic demands of those tissues. Early detection will be the key to limiting the effects on the body. Signs and symptoms of shock include the following:

- Apprehension, anxiety, restlessness and irritability.
- Altered level of consciousness.
- Nausea and vomiting.
- Pale, ashen or grayish, cool and moist skin.
- Rapid breathing.
- Excessive thirst.

2-60. Anticipate shock in all injured personnel. Treat all injured persons as follows, regardless of what symptoms appear:

- If victims are conscious, place them on a level surface with the lower extremities elevated 6 to 8 inches.
- If victims are unconscious, place them on their side or abdomen with their head turned to one side to prevent choking on vomit, blood, or other fluids.
- If unsure of the best position, place the victim perfectly flat. Once in a shock position, do not move the victim.
- Maintain body temperature by insulating the victim from the surroundings and, in some instances, applying external heat.
- If the victim's clothing is wet, remove it as soon as possible and replace with dry clothing.
- Improvise a shelter to insulate the victim from the weather.

- Use warm liquids or foods, a sleeping bag, another person, warmed water in canteens, hot rocks wrapped in clothing, or fires on either side of the victim to provide external warmth.
- If the victim is conscious, slowly administer small doses of a warm salt or sugar solution, if available.
- If the victim is unconscious or has abdominal wounds, do not give fluids by mouth.
- Have the victim rest for at least 24 hours.
- If the victim is a lone isolated person, the victim should lie in a depression in the ground, behind a tree, or any other place out of the weather, with the head lower than the feet.
- Reassess the victim constantly.

2-61. Assess shock in all victims as shown in Figure 2-4 on page 47.

Figure 2-4. Treatment of Shock

Conscious victim
- Place on level surface.
- Remove all wet clothing.
- Give warm fluids.
- Allow at least 24 hours rest.
- Insulate from ground.
- Shelter from weather.
- Maintain body heat.
- Elevate lower extremities 15~20 cm (6~8 inches).

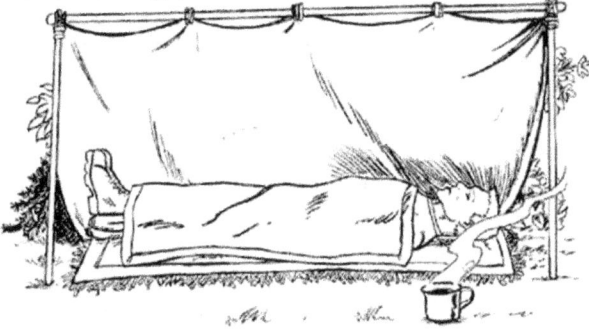

Unconscious victim
Same as for conscious victim except—
- Place victim on side and turn head to one side to prevent choking on vomit, blood, or other fluids.
- Do not elevate extremities.
- Do not administer fluids.

ILLNESS, INFECTION, SOFT TISSUE TRAUMA

2-62. Isolated persons will face debilitating illness, infection, and soft tissue trauma which can become life-threatening during isolation when they are without adequate medical care and basic personal hygiene capabilities.

2-63. Table 2-1 lists typical illnesses and infection, their signs and symptoms and treatment. Keep in mind that some of the treatments are extreme measures and only meant to be applied with extreme caution during actual isolation.

Table 2-1. Typical Illnesses, Infections, Causes and Treatments

Beriberi

Signs & Symptoms	Lack of vitamin B-1 (thiamine) induces the following symptoms: tiredness, irritation, nausea, constipation, poor memory, numb toes, burning sensation in the feet, sore legs, calf muscle cramps, unsteady walk, difficulty in getting up from a squat, weakened heart, heart failure, bloating.
Treatment	Replace thiamine by eating thiamine-rich foods including brown rice, whole grains, raw fruits & vegetables, legumes, seeds, nuts, yogurt, pork, beef, liver, brewer's yeast.

Body Lice

Signs & Symptoms	A small (1/16 inch) long insect that feeds on skin. Associated with wearing the same clothes for long periods without laundering, shared bedding, and close contact. Induces the following symptoms: Intense itching with deep scratches around shoulders, flanks, neck. Bites appear as small red pimples; may cause generalized skin rash.
Treatment	Washing clothing and bedding in boiling water, treating clothing with 1% malathion powder or 10% DDT powder.

Conjunctivitis (Pinkeye)

Signs & Symptoms	Viral and bacterial pinkeye are contagious and spread very easily. Symptoms include: swelling of the eyelids, itching or burning feeling of the eyelids, swollen and tender areas in front of the ears, a lot of tearing, clear or slightly thick, whitish drainage.

Treatment	Keep the eyes clear of discharge. Do not patch the eyes. Since most pinkeye is caused by viruses for which there is usually no medical treatment, preventing its spread is important. Wash hands thoroughly to avoid spreading the infection. Avoid sharing an object, such as a washcloth or towel, with a person who has pinkeye (which can spread the infection). Viral pinkeye symptoms usually last 5 to 7 days but may last up to 3 weeks and can become ongoing or chronic.

Dysentery / Diarrhea

Signs & Symptoms	60% of isolated personnel will experience diarrhea. Dysentery is a type of gastroenteritis that results in diarrhea with blood. Dysentery/Diarrhea is caused by infection with a parasitic amoeba or one of several bacteria types that induces the following symptoms: bloody diarrhea, fever, stomach pain, rapid weight loss. Can lead to death.
Treatment	Rehydration through drinking, use of an intravenous (IV) drip to replace lost fluids and antibiotics to control infection. Prevented through cleanliness, wastewater treatment, and personal hygiene using boiling water. Limit intake to fluids only for 24 hours. Drink 1 cup of a strong solution of tea every two hours until the diarrhea slows or stops. The tannic acid in the tea helps to control diarrhea. Make a solution of one handful of ground chalk, charcoal, or dried bones and a canteen cup of treated water. An equal portion of apple pomace (solid remains after pressing for juice) or the rinds of citrus fruit add to the mixture will make it more effective. Take 2 tablespoons of the solution every two hours until the diarrhea slows or stops.

Fever

Signs & Symptoms	A fever is a temporary increase in body temperature, often due to an illness. Fever is a sign that something out of the ordinary is going on in the body.
	For an adult, a fever may be uncomfortable, but usually isn't a cause for concern unless it reaches 103 F (39.4 C) or higher. For infants and toddlers, a slightly elevated temperature may indicate a serious infection.
	Additional fever signs and symptoms may include sweating, chills and shivering, headache, muscle aches, loss of appetite, irritability, dehydration, general weakness.
Treatment	Sponge off with cool water or alcohol. Use a fan. If aspirin is taken, the dosage is two tablets every six hours.
	Weeping willow bark is one of the most well-known sources of salicin (a naturally occurring antipyretic; see Paragraph 2-79). Soak the bark scrapings off several twigs and soak them in one cup of hot water for 10 minutes for a bitter tasting anti-diarrhea drink. Take a few sips every 2 hours, and continue until the symptoms subside. Drink plenty of fluids.

Food Poisoning

Signs & Symptoms	Bacterial contamination of food sources has historically caused much more difficulty in survival situations than the ingestion of poisonous plants and animals. Food poisoning strikes when you eat food that is contaminated with bacteria, viruses or parasites or another toxin that has natural poisonous properties. Signs and symptoms include: nausea, vomiting, watery or bloody diarrhea, abdominal pain and cramps, fever. Symptoms may start within hours after eating the contaminated food, or may begin days or even weeks later. Sickness caused by food poisoning generally lasts from a few hours to several days.
Treatment	Determine what made you sick and what you should do. Increase fluid intake to flush toxins from your body. Limit eating solid food. Replace electrolytes. Rest. Wash / sterilize your hands and cooking utensils. Properly store food. Cook meat thoroughly. Keep the casualty quiet, lying down, and drinking substantial quantities of water. Eat small amounts of fine, clean charcoal frequently. If detained or captive: If chalk is available, reduce it to powder, and eat to coat and soothe the intestines. Proper sanitation and personal hygiene will help prevent the spread of infection.

Hepatitis

Signs & Symptoms	Inflammation of the liver, called "hepatitis", is caused by various viruses which are harder to cure than bacteria (like that which causes the plague). There are several types of hepatitis. Water in the isolation environment often contains harmful microorganisms, bacteria and parasites that can cause a variety of ailments, such as giardia, dysentery, hepatitis, and hookworms. Contaminated food and/or water serve as the vehicle of transmission. Incubation runs 2~6 weeks, and the acute stage lasts 2~12 weeks with recovery being several weeks to several months.
Treatment	Force fluids; avoid strenuous exercises; abstain from alcohol; thoroughly wash hands and utensils, bedding, and clothing. Boil water. Disinfection of fecal material may be necessary, and increasing vitamin B intake will help. Medicinal herbs that are effective against hepatitis include coffee (Coffea arabica) at 500 mg per day (approx. 5 cups). Be careful not to take more than 1000 mg/day. Celandine (Chelidonium majus) is an herb that has aerial parts and roots that work against hepatitis. For a tea, use 1 ½ spoonfuls in boiling water and steep (allow to soak) for 10 minutes.

Pneumonia

Signs & Symptoms	With pneumonia you generally have all the symptoms of flu, but also: High fever up to 105 F. Coughing out greenish, yellow, or bloody mucus. Chills that make you shake. Feeling like you can't catch your breath, especially when you move around a lot. Feeling very tired. Low appetite. Sharp or stabbing chest pain (you might feel it more when you cough or take a deep breath). Sweating a lot. Fast breathing and heartbeat. Lips and fingernails turning blue. Confusion.
Treatment	Complete bed rest and aspirin. Willow provides aspirin-like relief. Pleurisy root consumed as a tea of boiled roots treats pneumonia and the expulsion of phlegm. Deep-breathing exercises may be helpful to promote coughing up phlegm. Drinking plenty of fluids will help to loosen the phlegm. Hydrogen peroxide was successfully used as a pneumonia treatment in the 1920s.

Tuberculosis (TB)

Signs & Symptoms	Caused by bacterial infection with the following symptoms: Coughing, initially bringing up yellow-green phlegm that becomes tinged with blood; low fever, chest pain, loss of appetite, increasingly severe lung symptoms, severe weight loss, a wasted appearance, can cause death.
Treatment	Treated by antibiotics, rest, and good nutrition. Lungs are sometimes collapsed surgically so they can rest and heal.

Worms and Intestinal Parasites

Signs & Symptoms	Internal parasites enter the body through contaminated food and water. Undercooked meat is a common place for parasites to hide, as well as water and associated plants from contaminated lakes, ponds, or creeks. Never eat raw vegetables contaminated by raw sewage. • Tapeworms can grow as long as 60 feet while living in the human intestine. There are currently more than 5,000 different species of tapeworm. • Hookworms generally suck blood from your intestinal walls. • Giardia is a single-celled parasite, infection by which is usually the result of drinking infected waters. It typically survives in chlorinated water and commonly lives in mountain streams. • Blood flukes mature first in snails and then complete their life cycle by burrowing through human skin and swimming through veins. Blood flukes infect more than 200 million people. There are also other species of flukes found living in the liver, lungs, and pancreas. • In the case of malaria, a new generation of parasitic microbugs will burst from a single red blood cell. • Pinworms are the most common roundworm. Worldwide, roughly 209 million people are infected. The most common sign of pinworm infestation is anal itching at night, which is when the female pinworm migrates to the perineum to lay her eggs.

Treatment	• Saltwater. Mix 4 tablespoons of salt into 1 quart of water. DO NOT repeat this treatment.
	• Tobacco. Eat 1 to 1 1/2 cigarettes or approximately 1 teaspoon (pinch) of smokeless tobacco (chewing/dipping tobacco, etc.) The nicotine in the tobacco will kill or stun the worms long enough for your system to pass them. If the infestation is severe, repeat the treatment in 24 to 48 hours, but no sooner.
	• Kerosene. Drink 2 tablespoons of kerosene, but no more. If necessary, you can repeat this treatment in 24 to 48 hours. Be careful not to inhale the fumes. They may cause lung irritation.
	• Hot peppers. Add hot peppers to soup, rice, and meat dishes or eat them raw. For this treatment to be effective, make peppers a steady part of the diet.
	• Garlic. Chop or crush four cloves, mix with one glass of liquid, and drink daily for 3 weeks.

IV — intravenous
DDT — Dichlorodiphenyltrichloroethane

TREATING INFECTION

2-64. During isolation, including detention or captivity, consider all breaks in the skin contaminated, and treat them appropriately. Clean even superficial scratches with soap and water and treat them with antiseptics, if available.

Basic Treatment

2-65. Open wounds are serious in a survival situation, not only because of tissue damage and blood loss, but also because they may become infected. Bacteria on the object that made the wound, on the individual's skin and clothing, or on other foreign material or dirt that touches the wound may cause infection. By taking proper care of the wound, you reduce the chance of further contamination and promote healing. Clean the wound fully as soon as possible by doing the following:

• Removing or cutting clothing away from the wound.
• Look for an exit wound if a sharp object, gunshot, or projectile caused a wound.
• Thoroughly clean the skin around the wound.

- Rinse (DO NOT SCRUB) the wound with large amounts of water under pressure. Water can be pressurized using a syringe, irrigator, or plastic baggie with a small hole poked in it. Water is the most universally available cleaning agent. Water used to cleanse a wound should be potable at a minimum, with sterile water preferred. At sea level, sterilize water by placing it in a covered container and boiling it for 10 minutes. Above 3,000 feet, water should be boiled for one hour (in a covered container) to ensure adequate sterilization. The water will remain sterile and can be stored indefinitely as long as it is covered.)

- Care must be taken not to rinse/irrigate the wound so vigorously that clots are washed away and bleeding resumes. Allow a period of an hour after the bleeding has been stopped before beginning irrigation with the sterile water. Begin gently at first, removing unhealthy tissue, increasing the vigor of the irrigation over a period of time, and avoid doing additional damage to the wound. The wound should be left open to promote cleansing and drainage of infection.

Open Treatment

2-66. The "open treatment" method is the safest way to manage wounds in survival situations. Leave the wound open to allow the drainage of any pus resulting from infection. As long as the wound can drain, it generally will not become life-threatening, regardless of how unpleasant it looks or smells. Cover the wound (including nerves, bone, and blood vessels) with a clean dressing. Place a bandage on the dressing to hold it in place. Change the dressing daily to check for infection.

2-67. Wounds, left open, heal by formation of infection resistant granulation tissue known as "proud flesh." This tissue is easily recognized by its moist red granular appearance, a good sign in any wound. A notable exception to "open treatment" is the early closure of facial wounds which interfere with breathing, eating, or drinking.

Gaping Wounds

2-68. If a wound is gaping, you can bring the edges together with adhesive tape cut in the form of a "butterfly" or "dumbbell". Use this method with extreme caution in the absence of antibiotics. You must always allow for proper drainage of the wound to avoid infection. Immobilization in a position to favor adequate circulation, both to and from the wound, will typically hasten the healing of major wounds/lacerations.

Infections in Wounds

2-69. In a survival situation, some degree of wound infection is almost inevitable. Pain, swelling, and redness around the wound, increased temperature, and pus in the wound or on the dressing indicate infection is present. If the wound becomes infected, treat it as follows:

- Place a warm, moist compress with lukewarm saltwater directly on the infected wound to help draw out infection and promote oozing of fluids from the wound, thereby removing toxic products.
- Apply a warm compress and change it when it cools. Keep a warm compress on the wound for a total of 30 minutes.
- Apply the warm compress three or four times daily.
- Drain the wound. Open and gently probe the infected wound with a sterile instrument.
- Dress and bandage the wound.
- Drink a lot of water.
- In the event of gunshot or other serious wounds, it may be better to rinse the wound out vigorously every day with the cleanest water available. If drinking water or methods to purify drinking water are limited, do not use your drinking water. Flush the wound forcefully daily until the wound is healed over.
- Continue this treatment daily until all signs of infection have disappeared.
- Apply heat to further aid in mobilizing local body defense measures. Soak the wound in lukewarm saltwater.

Use of Maggots

2-70. If you do not have antibiotics and the wound has become severely infected, does not heal, and ordinary debridement is impossible, consider maggot therapy. Maggots frequently infest deep open wounds. The natural tendency is to remove these maggots. But, they actually do a good job of cleansing a wound by removing dead tissue. Maggots exude calcium carbonate which alkalizes the wound and increases the destruction of bacteria and dead tissue by the body's white blood cells. A maggot's excretion also contains two chemicals (allantoin and urea) which stimulate growth of healthy tissue and hasten wound healing. Use the following process when using maggots.

- Expose the wound to flies for one day and then cover it.
- Check daily for maggots.
- Once maggots develop, keep wound covered but check daily.
- Remove all maggots when they have cleaned out all dead tissue and before they start on healthy tissue. Increased pain and bright red blood in the wound indicate that the maggots have reached healthy tissue.
- Flush the wound repeatedly with potable (minimum) sterile (preferred) water to remove the maggots.

- Bandage the wound and treat it as any other wound. It should heal normally.

Debridement

2-71. Debridement is the surgical removal of lacerated, devitalized, or contaminated tissue. The debridement of severe wounds may be necessary to minimize infection (particularly of the gas gangrene type) and to reduce septic (toxic) shock. In essence, debridement is the removal of foreign material and dead or dying tissue. The procedure requires skill and should be done by nonmedical personnel only in an emergency. If debridement is required, follow these general guidelines:

- Cut away dead skin.
- Muscle may be trimmed back to a point where bleeding starts and gross discoloration ceases.
- Cut away damaged fat.
- Conserve bone and nerves where possible and protect from further damage.
- Ample natural drainage for the potentially infected wound and final closure of the wound.

Suture

2-72. Despite the danger of infection, occasionally it may be necessary to suture a wound in order to control bleeding or increase the mobility of the patient. Procure thread from parachute lines, fabric, or clothing, and stitch the wound closed by "suturing." If suturing is required, place the stitches individually and far enough apart to permit drainage of underlying parts. Do not worry about the cosmetic effect; just suture the tissue together. For a scalp wound, use hair to close it. Infection is less a danger in this area due to the rich blood supply. Remember that in most situations, it is imperative that the wound be left open and allowed to drain.

Medicinal Plants

2-73. In using plants for medical treatment, positive identification of the plants involved is as critical as when using them for food. Proper use of these plants is equally important. Many natural remedies work slower than manufactured medicines. Therefore, start with smaller doses and allow more time for them to take effect.

WARNING: The following remedies are for use only in a survival situation. Do not use them routinely as some can be potentially toxic and have serious long-term effects.

2-74. The following terms and their definitions are associated with medicinal plant use:

- Poultice. This is crushed leaves or other plant parts, possibly heated, that are applied to a wound or sore either directly or wrapped in cloth or paper. Poultices, when hot, increase the circulation in the affected area and help healing through the chemicals present in the plants. As the poultice dries out, it draws the toxins out of a wound. A poultice should be prepared to a "mashed potatoes-like" consistency and be applied as warm as the patient can stand.

- Infusion or tisane[1] or tea. This blend is the preparation of medicinal herbs for internal or external application. Place a small quantity of an herb in a container, pour hot water over it, and let it steep (covered or uncovered) before use. Care must always be taken to not drink too much of a tea in the beginning of treatment as it may have adverse reactions on an empty stomach.

- Decoction. This is the extract of a boiled-down or simmered herb leaf or root. Add herb leaf or root to water. Bring them to a sustained boil or simmer them to draw their chemicals into the water. The average ratio is about 1 to 2 ounces (28 to 56 grams) of herb to 0.5 liter of water.

- Expressed juice. These are liquids or saps squeezed from plant material and either applied to the wound or made into another medicine.

SURVIVAL MEDICINE PROPERTIES

2-75. Survival medicine is practiced by isolated personnel to aid in their survival during isolation.

TANNIN / TANNIC ACID

2-76. Prevents or reduces the effects of dysentery/diarrhea (anti-diarrheal).

- Medicinal uses. Diarrhea, dysentery, burns, skin problems, and parasites. Tannin/tannic acid solution prevents infection and aids healing.

- Sources. All thread plants contain tannic acid and it is found in the outer bark of all hardwood trees, acorns, banana plants, common plantain, strawberry leaves, blackberry stems, roots of blackberries. White oak bark and other barks containing tannin are

---

1. Tisane: a tea-like infusion made from material derived from plants other than the tea plant (Camellia sinensis) — commonly known as "herbal tea."

also effective when made into a strong tea (including cowberry, cranberry, or hazel leaves).

Preparation

2-77. Tannic acid is also found in the moist inner bark of hardwood trees. Boil the inner bark of a hardwood tree for two hours or more to release the tannic acid. Although this solution has a vile taste and smell, it will stop most cases of diarrhea.

- Place crushed outer bark, acorns, or leaves in water.
- Leach out the tannin by soaking or boiling.
- Increase tannin content by longer soaking time.
- Replace depleted material with fresh bark/plants.
- Clay, ashes, charcoal, powdered chalk, powdered bones, and pectin can be consumed or mixed in a tannic acid tea in a crude form of Kaopectate[1] with good results.

Treatment

2-78. Burns.

- Moisten bandage with cooled tannin tea.
- Apply compress to burned area.
- Pour cooled tea on burned areas to ease pain.
- Diarrhea, dysentery, and worms. Drink strong tea solution (may promote voiding of worms).
- Skin problems (dry rashes and fungal infections). Apply cool compresses or soak affected part to relieve itching and promote healing.
- Lice and insect bites. Wash affected areas with tea to ease itching.

SALICIN / SALICYLIC ACID

2-79. Prevents or reduces the effects of fever (antipyretic).

- Medicinal uses. Aches, colds, fever, inflammation, pain, sprains, and sore throat are relieved through its aspirin-like qualities.
- Sources. The weeping willow tree is a native of China and extra-tropical Asia. It belongs to the family of crack willows. This tree's bark is utilized in medicinal and tanning processes. Treat a fever with a tea made from willow bark, an infusion of elder flowers or fruit, linden flower, and aspen or slippery elm bark. Yarrow and peppermint teas are also very helpful.

---

1. Kaopectate (bismuth subsalicylate): an antacid medication used to treat gastrointestinal and stomach conditions. Commonly known by the commercial brand name Pepto-Bismol.

Preparation

- Gather twigs, buds, or cambium layer (soft, moist layer between the outer bark and the wood) of willow or aspen.
- Prepare tea.

2-80. Make a poultice:

- Crush the plant or stems.
- Make a pulpy mass.

Treatment

- Chew on twigs, buds, or cambium (the layer between the wood and the bark) for symptom relief.
- Drink tea for colds and sore throat.
- Use warm, moist poultice for aches and sprains.
    - Apply pulpy mass over injury.
    - Hold in place with a dressing.

PLANTAIN / WOUNDWORT / COMMON YARROW

2-81. Prevents or reduces the effects of hemorrhage and bleeding (including anti-hemorrhagic, anti-microbial, tissue regeneration, pain relief).

- Medicinal uses. A poultice of the leaves from common plantain, woundwort or common yarrow can be applied to wounds, stings, and sores in order to facilitate healing and prevent infection. The active chemical constituents in common plantain include aucubin (an anti-microbial agent), allantoin (which stimulates cellular growth and tissue regeneration), and mucilage (which reduces pain and discomfort). Plantain has astringent properties, and a tea made from the leaves can be ingested to treat diarrhea and soothe raw internal membranes as well as itching, wounds, abrasions, stings, diarrhea, and dysentery.
- Source. Common plantain is a species of flowering plant in the plantain family Plantaginaceae. The plant is native to most of Europe and northern and central Asia including Afghanistan, Pakistan and Iran and is widely naturalized throughout the world. There are over 200 plantain species with similar medicinal properties and it is one of the most abundant and widely distributed medicinal crops in the world. Additional sources include broadleaf plantain which is also a highly nutritious wild edible plant that is high in calcium and vitamins A, C, and K. The young, tender leaves can be eaten raw, and the older, stringier leaves can be boiled in stews and eaten. Prickly pear (the raw, peeled part) or witch hazel can be applied to wounds based upon their astringent properties (they shrink blood vessels). For bleeding

gums or mouth sores, sweet gum can be chewed or used as a toothpick.

Preparation
- Brew tea from seeds.
- Brew tea from leaves.
- Make poultice of leaves.

Treatment
- Drink tea made from seeds for diarrhea or dysentery.
- Drink tea made from leaves for vitamin and minerals.
- Use poultice to treat cuts, sores, burns, and stings.

TANSY / WILD CARROT / QUEEN ANNE'S LACE
2-82. Prevents or reduces the effects of internal parasites and worms (antihelminthics).

WARNING: Most treatment for worms or parasites are toxic. Therefore, all treatments should be used in moderation. DO NOT confuse Wild Carrot or Queen Anne's Lace with any of the Poison Hemlock family. Wild carrots and Queen Anne's Lace have hairy stems and stalk which is a very important identification element and separates the two carrots from several very deadly plants including Poison Hemlock (has smooth hollow stalks with purple blotches and no hairs on its stems), Water Hemlock and Fools Parsley. (Poison Hemlock, Water Hemlock and Fools Parsley smell nasty - just roll some leaves between your thumb and forefinger, and smell), while Wild Carrot roots smell like carrots.

- Medicinal uses. The wild carrot is an aromatic herb that acts as a diuretic, soothes the digestive tract, stimulates the uterus, supports the liver, stimulates the flow of urine and the removal of waste by the kidneys. A tea made from the roots is diuretic and has been used in the treatment of urinary stones. Very strong tannic acid can also be used with caution as it is very hard on the liver.
- Source. Tansy, Wild Carrot, Queen Anne's Lace.

Preparation

- Boil the leaves to make the infusion and remove any toxic elements. An infusion of the leaves has been used to counter cystitis and kidney stone formation, and to diminish stones that have already formed. A warm water infusion of the flowers has been used in the treatment of diabetes.

Treatment

- The grated raw root, especially of the cultivated forms, is used as a remedy for threadworms. The root is also used to encourage delayed menstruation, and can induce uterine contractions (should not be used by pregnant women).
- Carrots possess strong antiseptic qualities, can be used as a laxative, vermicide (worm-expelling agent), poultice and for the treatment of liver conditions.

YARROW / CATTAIL / WILD ONION / GARLIC / CHICKWEED / BURDOCK

2-83. Prevent or reduce the growth of disease/infection (antiseptic).

- Medicinal uses. Use antiseptics to cleanse wounds, snake bites, sores, boils, inflammations, burns or rashes. You can make antiseptics from the expressed juice of wild onion or garlic, chickweed leaves, or the crushed leaves of dock.
- Source. Yarrow is a flowering plant in the family Asteraceae. It is native to temperate regions of the Northern Hemisphere in Asia, Europe, and North America. The above-ground parts are used to make medicines to treat fever, common cold, hay fever, absence of menstruation, dysentery, diarrhea, loss of appetite, gastrointestinal (GI) tract discomfort, and to induce sweating. Some people chew the fresh leaves to relieve toothache. Other sources include Cattail, wild onion or garlic, chickweed, burdock root, mallow leaves or roots, white oak bark (tannic acid). Prickly pear, slippery elm, and sweet gum.

Preparation

- Two of the best antiseptics are sugar and honey.
  - Sugar should be applied to the wound until it becomes syrupy, then washed off and reapplied. Honey should be applied three times daily.
  - Honey is by far the best of the antiseptics for open wounds and burns, with sugar being second.

Treatment

- Apply poultice to affected area to stop bleeding.

- Yarrow is dried, powdered, and mixed with plantain or comfrey water to treat wounds or used fresh by itself.
- To treat burns, scrapes, insect bites and bruises, split open a cattail root and "bruise" the exposed portion so it can be used as a poultice that can be secured over the injured area
- The cattail stem has an amber- or honey-like substance that seeps from the stem. Use this secretion to treat small wounds and even toothaches, because it also has antiseptic properties.

POISONOUS PLANTS

2-84. Isolated persons can benefit greatly from using plants as a food source, medical aid, and shelter building material; however, successful use of plants in a survival situation depends on positive identification. Knowing poisonous plants is as important as knowing edible plants. Knowing poisonous plants will help isolated personnel avoid sustaining injuries from them. Plants generally poison by—

- Contact. With a poisonous plant causes any type of skin irritation or dermatitis.
- Ingestion. This occurs when a person eats a part of a poisonous plant.
- Absorption or inhalation. This happens when a person either absorbs the poison through the skin or inhales it into the respiratory system.

2-85. Isolated persons must understand that the effects of poisoning from plants range from minor irritation to death. It is difficult to say how poisonous a plant is because—

- Some plants require a large amount of contact before any adverse reaction occurs, while others will cause death with only a small amount of contact.
- Every plant will vary in the quantity toxins it contains due to different growing conditions and slight variations in subspecies.
- Every person has a different level of resistance to toxic substances.
- Some persons may be more sensitive to a particular plant.

2-86. The following are some common misconceptions about poisonous plants:

- "Watch the animals and eat what they eat." Most of the time this statement is true, but some animals can eat plants that are poisonous to humans.
- "Boil the plant in water and any poisons will be removed." Boiling removes many poisons, but not all.
- "Plants with a red color are poisonous." Some plants that are red are poisonous, but not all.

WARNING: Many poisonous plants look like their edible relatives or like other edible plants. For example, poison hemlock appears very similar to wild carrot.

2-87. Certain plants are safe to eat in certain seasons or stages of growth but poisonous in other stages. For example, the leaves of the pokeweed are edible when it first starts to grow, but they soon become poisonous. Some plants and their fruits are safe to eat only when they are ripe. For example, the ripe fruit of the May apple is edible, but all other parts and the green fruit are poisonous. Some plants contain both edible and poisonous parts; potatoes and tomatoes are common plant foods, but their green parts are poisonous. Some plants become toxic after wilting. For example, when the black cherry starts to wilt, hydrolyzing acid develops. Specific preparation methods make some plants edible that are poisonous raw. The thinly sliced and thoroughly dried (drying may take a year) corms of the jack-in-the-pulpit are safe to eat, but they are poisonous if not thoroughly dried. Learn to identify and use plants before you find yourself in a survival situation.

RULES FOR AVOIDING POISONOUS PLANTS

2-88. If isolated persons have little or no knowledge of the local vegetation, they should use the rules to select plants for the Universal Edibility Test. Remember to avoid all mushrooms. Mushrooms identification is very difficult and must be precise-even more so than with other plants. Some mushrooms cause death very quickly, and some mushrooms have no known antidote. Two general types of mushroom poisoning are gastrointestinal and central nervous system. Avoid contact with or touching unidentified plants unnecessarily.

AVOIDING CONTACT DERMATITIS

2-89. Contact dermatitis from plants will usually cause the most trouble for an isolated person. The effects may be persistent, spread by scratching, and be particularly dangerous if there is contact in or around the eyes. The principal toxin of these plants is usually oil that gets on the skin upon contact with the plant. The oil can also get on equipment and then infect whoever touches the equipment. Never burn a contact poisonous plant, because the smoke may be as harmful as the plant. You have a greater danger of being affected when you are overheated and sweating. The infection may be local or it may spread over the body. Symptoms may take from a few hours to several days to appear. Symptoms can include burning, reddening, itching, swelling, and blisters.

2-90. When you first have contact with the poisonous plants or when the first symptoms appear, you should try to remove the oil by washing with soap and cold water. If water is not available, you should wipe your skin repeatedly with dirt or sand. You should not use dirt if you have blisters (the dirt may break open the blisters and leave the body open to infection). After the removal of the oil, the area should be dried. The area can be washed with a tannic acid solution and jewelweed can be crushed and rubbed on the affected area to treat plant-caused rashes. Tannic acid can make from oak bark. Poisonous plants that cause contact dermatitis are—

- Cowhage.
- Poison ivy.
- Poison oak.
- Poison sumac.
- Rengas tree.
- Trumpet vine.

AVOIDING INGESTION POISONING

2-91. Ingestion poisoning can be very serious and could lead to death very quickly. You should not eat any plant unless you have positively identified it first. Isolated persons should keep a log of all plants eaten and plants identified as inedible if tested.

2-92. Symptoms of ingestion poisoning can include nausea, vomiting, diarrhea, abdominal cramps, depressed heartbeat and respiration, headaches, hallucinations, dry mouth, unconsciousness, coma; which can lead to death. If plant poisoning is suspected, try to remove the poisonous material from the victim's mouth and stomach as soon as possible. If the victim is conscious, induce vomiting by tickling the back of the victim's throat or by giving the victim warm saltwater. If the victim is conscious, dilute the poison by administering large quantities of water or milk.

2-93. The following plants can cause ingestion poisoning if eaten:

- Castor bean.
- Chinaberry.
- Death camas.
- Lantana.
- Manchineel.
- Oleander.
- Pangi.
- Physic nut.
- Poison and water hemlocks.
- Rosary pea.

- Strychnine tree.

TREATING INJURIES, BITES AND STINGS, POISONING, AND OTHER CONDITIONS

SOFT TISSUE TRAUMA

2-94. Soldiers face many threats in hostile fire environments. Whether conducting large-scale combat operations, low-intensity conflicts, or operations other than war; isolated personnel may wait for hours, days, weeks or longer before professional healthcare is available. It is important to understand the implications over the long run, of trauma and the potential for infection if not treated and properly cared for.

Blisters

2-95. Blisters are caused by friction and occur in areas where the epidermis is thick and tough enough to resist abrasion. At first there is a red, sore area called a hotspot. If the friction continues, the epidermis separates and fluid enters the space, causing a blister. The best treatment for blisters is to prevent them: make sure shoes/boots are properly fitted, keep your feet dry, wear two pairs of socks, check feet periodically, and stop at the first sign of rubbing. Apply a small piece of moleskin or athletic tape to areas that are causing problems.

2-96. The primary goal with blister treatment is to optimize comfort and continued activity, maximize prevention of infection, and prevent further blister enlargement. If a blister is smaller than a nickel, do not puncture it. Cut a doughnut-shaped piece of moleskin and center it over the blister. The doughnut hole prevents the adhesive from sticking to the tender blister and ripping it away when the moleskin is changed.

2-97. If the blister is larger than a nickel, it is best to drain it. Drain the blister by washing your hands and the area around the blister to decrease the chance of infection. Begin by puncturing the blister with a small sterile object, such as a needle or a pin, to drain the blister of built-up fluid. The best covering for a blister to heal is the skin itself. Clean and wash the blister and cover it with a sterile dressing.

Boils

2-98. A boil is a localized infection in the skin that begins as a reddened, tender area. Over time, the area becomes firm, hard, and increasingly tender. Eventually, the center of the boil softens and becomes filled with infection-fighting white blood cells from the bloodstream to eradicate the infection. This collection of white blood cells, bacteria, and proteins is known as pus. Finally, the pus forms a head which can be opened or may spontaneously drain out through the surface of the skin. Pus enclosed within tissue is referred to as an abscess. A boil is also referred to as a skin abscess. Boils can occur anywhere on the body, including the trunk, extremities, buttocks, groin, or other areas. The signs and symptoms of boils include the following:

- A firm, reddened bump.
- Tender, swollen skin surrounding the bump.
- A bump that may increase in size.
- Bump forms a pus-filled head which may spontaneously drain, weep, or ooze.

Treatment for boils includes the application of warm compresses to bring the boil to a head or use of the bottle suction method. Use the following process to perform the bottle suction method:

- Use an empty bottle that has been boiled in water.
- Place the opening of the bottle over the boil and seal the skin forming an airtight environment that will create a vacuum. This method will draw the pus to the skin surface when applied correctly.
- Open the boil using a sterile knife, wire, needle, or similar item.
- Thoroughly clean out the pus using soap and water.
- Cover the boil site, checking it periodically to ensure no further infection develops. Toothpaste and when available, iodine, were used to cover the boil site during the Vietnam War.

Boil a palmful of green leaves in one cup of water for 10 minutes to create an astringent. Soak a clean cloth in this brew and apply it directly to boils, abscesses, carbuncles, and ulcers when no other medical treatment is available.

Fungal Infections

2-99. Keep the skin clean and dry, and expose the infected area to as much sunlight as possible. Do not scratch the affected area. Antifungal powders, lye soap, chlorine bleach, alcohol, vinegar, concentrated saltwater, and iodine were used by Soldiers to treat fungal infections during the Vietnam War.

Facial Injury

2-100. Facial injuries are some of the most annoying and painful injuries—particularly injuries to the teeth and gums. Personnel should continue to clean and brush their teeth while isolated to prevent tooth decay and gingivitis. If a toothbrush is unavailable, make a chew stick by taking a small green stick, chewing the end, and using the fibers as a brush to clean the teeth. If tooth pain occurs, rinsing the mouth out with warm saltwater helps to soothe the pain. Personnel may also experience a painful mouth abscess. They can ease the pain and decrease the threat of infection by lancing the abscess and letting it drain. Other facial injuries may be soft tissue facial injuries such as abrasions, contusions, lacerations, and avulsions[1]. Clean and cover the facial injury several times a day, which will help prevent infection and crusting over.

Eye Injury

2-101. If an injury to the eyes occurs, you should know how to treat the injury to prevent loss of sight.

2-102. Snow blindness occurs when a person spends time in an area where bright sunrays reflect off the ice crystals in the snow. The eyes will become red, scratchy, and watery. The casualty may have headaches and eye pain with light exposure. The treatment for snow blindness is to cover the eyes with a dark cloth until the symptoms disappear. To prevent snow blindness, ultraviolet (UV) protection rated sunglasses should be worn or improvise a pair of sunshades for the eyes. Putting soot under the eyes helps reduce the shine and glare as well. This blinding affect can also occur when the sun's rays reflect off water and other surfaces; the treatment and prevention methods remain the same.

2-103. Should a person incur an eye impalement; never remove any object from the eye unless it is necessary to assist in their survival. Use an eye shield to cover the eye from further injury or if the eye shield is not available, make a protective cover from a cup, a bandage, or a similar item. This covering will protect the eye from further injury or movement. Cover both eyes because the eyes are sympathetic, and when one eye moves the other moves at the same time.

2-104. If chemicals or small objects get into the eye, flush the eyes with large amounts of water from inside the eye to the outside so the chemicals, etc. are not transferred from one eye and into the other.

BONE AND JOINT TRAUMA

2-105. The majority of isolated personnel will experience some form of bone or joint trauma during isolation. Broken bones are a serious survival emergency, removing the use of a limb and raising the risk of post-fracture infection. Additionally, the long term effects of detention and captivity highlight malnutrition and starvation's adverse impacts on bone health, bone quality, bone regeneration and decreased cortical[2] strength.

---

1. Avulsion: an injury in which a bodily structure is torn away by physical trauma.
2. Cortical bone is the hard, dense outer layer of bone.

Sprains

2-106. A sprain is an injury to a tendon or ligament in the body. Signs and symptoms associated with sprains include pain, swelling, bruising, and loss of functional movement. There may be a pop or a tear associated with the injury when it occurs. The treatment for sprains and strains are similar. When treating sprains, you should follow the letters in RICE.

2-107. RICE as defined—

- **R-Rest** injured area.
- **I-Ice** for 24 to 48 hours. Use improvised ice packs and cold compresses intermittently for up to forty-eight hours following the injury.
- **C-Compression** wrap or splint to help stabilize. If possible, leave the boot on a sprained ankle unless circulation is compromised.
- **E-Elevate** the affected area.

Strains

2-108. Strains are an injury to the muscle or the tendon, the tissue that connects the muscle to the bone. Signs and symptoms of a strain typically include pain, muscle spasms, and muscle weakness. There can be swelling, cramping, inflammation and, in severe cases, some loss of muscle functions. Severe strains that partially or completely tear the muscle or the tendon can be very painful and disabling. Treatment follows the RICE algorithm used to treat sprains.

Fractures and Dislocations

2-109. Combat, isolating events, and accidents produce broken bones and compression fractures affecting the back. Broken bones should be treated early, before swelling complicates treatment. Perform the MARCH algorithm conditions first (see page 29). Immobilize before moving, if possible, and finish treatment later.

2-110. There are basically two types of fractures: open and closed.

2-111. With an open (or compound) fracture, the bone protrudes through the skin and complicates the actual fracture with an open wound. In open fractures, the soft tissue injury must be treated and properly dressed before splinting. Any bone protruding from the wound should be cleaned with an antiseptic and kept moist. Splint the injured area and continually monitor blood flow past the injury. Only reposition the break if there is no blood flow.

2-112. The closed fracture has no open wounds. The signs and symptoms of a closed fracture are pain, tenderness, discoloration, swelling deformity, loss of function, and grating (a sound or feeling that occurs when broken bone ends rub together). Under normal conditions, the treatment of a fracture is beyond the abilities of normal first aid. However, in survival medicine, the risk of potential bone deformities is outweighed by the necessity to hasten the healing process and return the limb to function to enable the return of the isolated person to safety.

2-113. Dangers associated with the treatment of a fracture include severing or compression of a nerve or blood vessel at the site of fracture. For this reason minimum manipulation should be done, and only very cautiously. If the area below the break becomes numb, swollen, cool to the touch, or turning pale, and the casualty showing signs of shock, a major vessel may have been severed and this internal bleeding must be controlled. Reset the fracture and treat the casualty for shock and replace lost fluids.

2-114. Often during the treatment of fractures, traction (i.e. a slow, strong pull — not a tug) must be applied during the splinting process to facilitate healing. Fractures of the smaller bones (finger or hand) can usually be performed using hand traction. For larger bones and muscled areas, traction is applied by wedging the fractured leg or arm in the V-notch of a tree and pulling until the overriding edges of fractured bone are brought into line. Check the alignment against the other limb. Then splint and immobilize, keeping up the traction to ensure it does not slip back. A splint can be made from sticks, branches, parts of wreckage, driftwood, rolls of paper, and other rigid items.

2-115. Use field-expedient methods to immobilize a fracture. (See Figure 2-5 on page 72). Once an injury is immobilized, assess the pulse to ensure proper blood flow to the injury. Take your pulse multiple times at the farthest part of the injury. When immobilizing a fracture, it is crucial to splint the joints above and below the fracture site to prevent the bone ends from moving. If the mid forearm is fractured, the wrist and the elbow should be splinted to keep the bone ends from moving and achieve proper healing. Immobilize a joint in a fixed neutral functional position. This position is neither completely straight nor completely flexed, but in a position about midway between. Splint the finger, to about the same position that the finger would attain at rest. Use large straight strips of cloth material, vines, or braided plant material to secure the splints. To aid in controlling swelling, the limb should be elevated repeatedly throughout the day.

Figure 2-5. Building a Splint

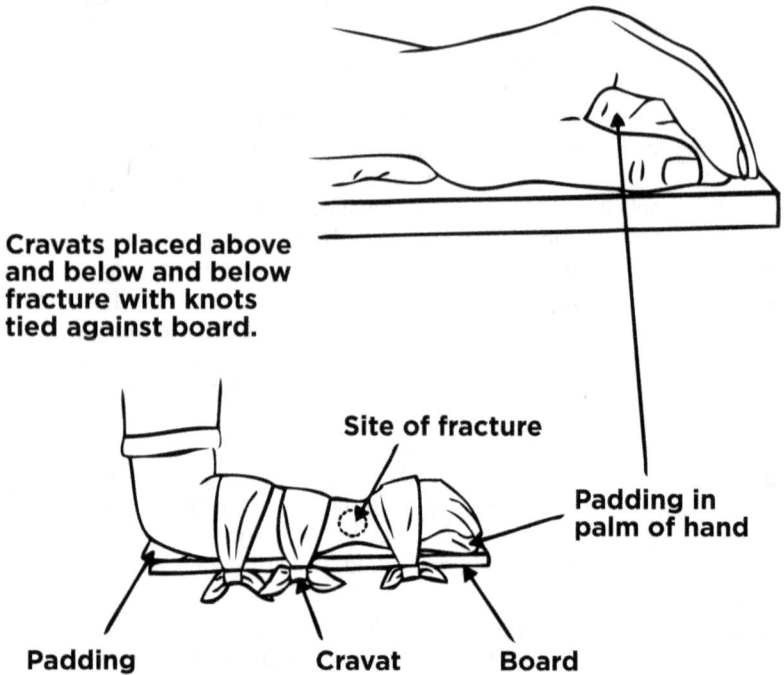

Cravats placed above and below and below fracture with knots tied against board.

Site of fracture

Padding in palm of hand

Padding   **Cravat**   **Board**

2-116. Improperly aligned bones and separation of bone joints are called a dislocation. These misalignments can be extremely painful and can cause impairment of nerve or circulatory function below the area affected. An isolated person must place these joints back into alignment as quickly as possible. Signs and symptoms of dislocations are joint pain, tenderness, swelling, discoloration, limited range of motion, and joint deformity. Manual traction or the use of weight (body weight) to pull the bones is the safest and easiest for "resetting" a dislocation and allows for normal function and circulation. Without an X-ray, judge the proper alignment by the look and feel of the joint and by comparing it to the joint on the opposite side. Immobilization is nothing more than splinting the dislocation after resetting. Figure 2-6 on page 73 demonstrates the use of field-expedient devices for an improvised sling.

Figure 2-6. Slings

Shirt tail use   **Belt use**

Improvised sling

INSECT BITES AND STINGS

2-117. Insects and related pests are hazards in a survival situation. They cause irritations and are often carriers of diseases that cause severe allergic reactions in some individuals. In many parts of the world, people can be exposed to serious, even fatal, diseases not encountered in the United States. Common insects and their hazards include the following:

2-118. Ticks. Ticks can carry and transmit diseases, such as Rocky Mountain spotted fever, common in many parts of the United States. They also transmit Lyme disease. Take care to remove the whole tick. Use tweezers if available. Grasp the tick where the mouthparts are attached to the skin and pull up with steady, even pressure. Do not twist or jerk the tick. Do not squeeze the tick's body. Wash hands after touching the tick. Clean the tick wound daily until healed.

2-119. Mosquitoes. Mosquitoes may carry malaria, dengue, and many other diseases.

2-120. Flies. Flies can spread disease from contact with infectious sources. They are carriers of sleeping sickness, typhoid, cholera, and dysentery.

2-121. Fleas. Fleas can transmit plague.

2-122. Lice. Lice can transmit typhus and relapsing fever.

2-123. The best method of preventing bites and stings is to be aware of your environment and the insects in the area. If insects cannot be avoided, wrap your face with a bandana, use netting, roll down your sleeves, wear gloves, and use insect repellent.

2-124. If bitten or stung, do not scratch the bite or sting; it might become infected. Personnel should inspect their body at least once a day to ensure there are no insects attached to them. It is impossible to list the treatment for all the different types of bites and stings. However, bites and stings can generally be treated as follows.

2-125. Become familiar with available antibiotics and how to use them before deployment.

2-126. Make sure immunizations are up-to-date before deployment; this can prevent most common diseases carried by mosquitoes and some carried by flies.

2-127. Be aware that common fly-borne diseases are usually treatable with penicillin or erythromycin.

2-128. Be aware that most tick, flea, louse, and mite-borne diseases are treatable with tetracycline.

2-129. Be aware that most antibiotics come in 250-milligram or 500-milligram tablets. If you cannot remember the exact dose rate to treat a disease, two tablets, four times a day for 10 to 14 days will usually kill any bacteria.

SNAKEBITES

2-130. Deaths from snakebites are rare. More than one-half of snakebite victims have little or no poisoning, and only about one-quarter develop serious systemic poisoning. The primary concern in the treatment of snakebites is to limit the amount of eventual tissue destruction around the bite area. A bite wound, regardless of the type of animal that inflicted it, can become infected from bacteria in the animal's mouth. With venomous and nonvenomous snakebites, this local infection is responsible for a large part of the residual damage that results.

2-131. Snake venoms contain poisons that attack the victim's central nervous system (neurotoxins) and blood circulation (hemotoxins) as well as poisonous digestive enzymes (cytotoxins) that aid in digesting their prey. These poisons can cause a very large area of tissue death, leaving a large, open wound. This condition could lead to the need for eventual amputation if not treated. Shock and panic in a snakebite victim can also affect the person's recovery. Excitement, hysteria, and panic can speed up circulation, causing the body to absorb the toxin quickly. Signs of shock occur within the first 30 minutes after a bite.

2-132. Before isolated persons start treating a snakebite, they should determine whether the snake was venomous or nonvenomous. Bites from a nonvenomous snake will show rows of teeth marks. Bites from a venomous snake may have rows of teeth marks showing, but will have one or more distinctive puncture marks caused by fang penetration. Symptoms of a venomous bite may be spontaneous bleeding from the nose and anus, blood in the urine, pain at the site of the bite, and swelling at the site of the bite within a few minutes or up to 2 hours later. Breathing difficulty, paralysis, weakness, twitching, and numbness are also signs of neurotoxic venoms. These signs usually appear 1.5 to 2 hours after the bite. If personnel determine that a venomous snake bit an individual, use the following procedures:

- Keep the victim still and reassured.
- Set up for shock and force fluids or give by intravenous (IV) means.
- Remove watches, rings, bracelets, or other constricting items.
- Clean the bite area.
- Maintain an airway (especially if bitten near the face or neck) and be prepared to administer mouth-to-mouth resuscitation or cardio-pulmonary resuscitation (CPR).
- Use a constricting band between the wound and the heart.
- Immobilize the site.
- Get the person to medical treatment as soon as possible.
- You should know four very important guidelines during the treatment of snakebites.
  - **Do not** give the victim or take alcoholic beverages or tobacco products, or administer atropine, morphine or other central nervous system depressors.
  - **Do not** make any deep cuts at the bite site. Cutting opens capillaries that in turn open a direct route into the blood stream for venom and infection.
  - **Do not** put the hands on the face or rub the eyes, as venom may be on hands. Venom may cause blindness.
  - **Do not** break open the large blisters that form around the bite site.
- After caring for the victim as described, take the following actions to minimize local effects:
  - If infection appears, **do** keep the wound open and clean.
  - **Do** use heat after 24 to 48 hours to help prevent the spread of local infection. Heat also helps to draw out an infection.

- **Do** keep the wound covered with a dry, sterile dressing.
- **Do** have the victim drink large amounts of fluids until the infection is gone.
- When treating a snakebite, **do** be sure to keep the extremity lower than the heart.

ENVIRONMENTAL INJURY

2-133. The successful prevention and control of cold, heat and altitude injuries depend on vigorous command interest, the provision of adequate clothing, and a number of individual and group measures.

High Altitude Sickness

2-134. Persons not acclimated to high altitude may experience several illnesses that could affect their body and require treatment. Rapid change of altitude—for example, going from sea level to high altitude—can cause loss of consciousness followed shortly by death. The typical person working in a high-altitude environment will need to acclimatize in order to prepare their body for functioning at altitude. Acclimatization is the process of physiological adjustments made by the body's oxygen delivery system to the cells; this improves their hypoxic tolerance.

2-135. The first and maybe only sign of high-altitude illness is a headache. A high-altitude headache will have at least two of the following characteristics: bilateral, frontal, or frontal temporal; dull or pressing quality; mild or moderate in intensity; and aggravated by exertion, movement, straining, coughing, or bending. Acute mountain sickness symptoms include headache plus one or more of the following: nausea, vomiting, fatigue, dizziness, and difficulty sleeping.

High-altitude Pulmonary Edema

2-136. High-altitude pulmonary edema (HAPE) is one of the most severe high-altitude illnesses. This illness can be so severe it can cause death. Treatment for high-altitude pulmonary edema involves limiting exertion, staying warm, and rapidly descending from altitude. Signs of high-altitude pulmonary edema onset involve loss of appetite, low energy level, coughing spasms, and a headache.

2-137. High-altitude cerebral edema (HACE) is another serious risk at high altitudes. It is a severe high-altitude illness where the brain swells and stops functioning. This illness is so severe that it can cause death. Other than rest, hydration and descending to a lower altitude, further treatment is beyond the capabilities of nonmedical personnel.

Note: At high altitudes, the body will require more calories to counteract weight loss and acute loss of appetite.

Heat Injuries

2-138. In a very warm climate, personnel may experience heat injuries. Heat injuries can occur suddenly and, if not treated properly, can quickly incapacitate and prevent mission completion. Personnel should limit their time in the hot sun, cover exposed skin or wear sunscreen, remain hydrated, and take plenty of rest breaks. Follow this formula, otherwise heat cramps, heat exhaustion, and heatstroke may occur.

2-139. Heat cramps are muscle cramps in the arms, legs, and/or stomach. Other symptoms experienced may be wet skin and extreme thirst. The treatment for heat cramps involves resting in a cool, shaded area and loosening restrictive clothing. Personnel should slowly sip one quart of water per hour. Add one-fourth teaspoon salt to the water to help replace minerals lost through sweating.

2-140. If untreated, heat cramps can progress into heat exhaustion. With heat exhaustion, personnel may experience cool, moist, and clammy skin; a loss of appetite; a headache; excessive sweating; and weakness. They may also experience faintness, dizziness, nausea, and muscle cramps. The treatment for heat exhaustion involves resting in a cool, shaded area and loosening restrictive clothing. Personnel should pour cool water over the person and fan to increase the cooling effect of evaporation. They should also slowly sip at least one quart of water per hour to replace lost fluids. Elevating the legs will help.

2-141. Heatstroke is the most dangerous heat injury. The casualty will stop sweating and have hot, dry skin. They may experience a headache, dizziness, nausea, vomiting, and a rapid pulse and respiration. They may have seizures and mental confusion, and may collapse and lose consciousness. The treatment for heatstroke involves rapidly cooling victims by immersing them in cool water. If a water bath is unavailable, then douse victims clothing with water making sure to cool the groin, the armpit, and the back and neck areas. Use caution to avoid over-cooling the victims. If victims are conscious, have them consume one quart of water.

Cold Injuries

2-142. In cold environments, personnel may succumb to different types of cold injuries that can cause incapacitation and permanent damage and make it difficult or impossible to reach mission success. To reduce the possibility of cold injuries, limit your time in the extreme cold, cover exposed skin, and use the buddy system to check for cold injuries. Proper layering of clothing along with staying dry and hydrated with warm fluids can reduce the risk of cold injuries. Heat transfers to and from the body through radiation, convection, conduction, evaporation, and respiration (see Figure 2-7 on page 78).

Figure 2-7. Heat Transfer

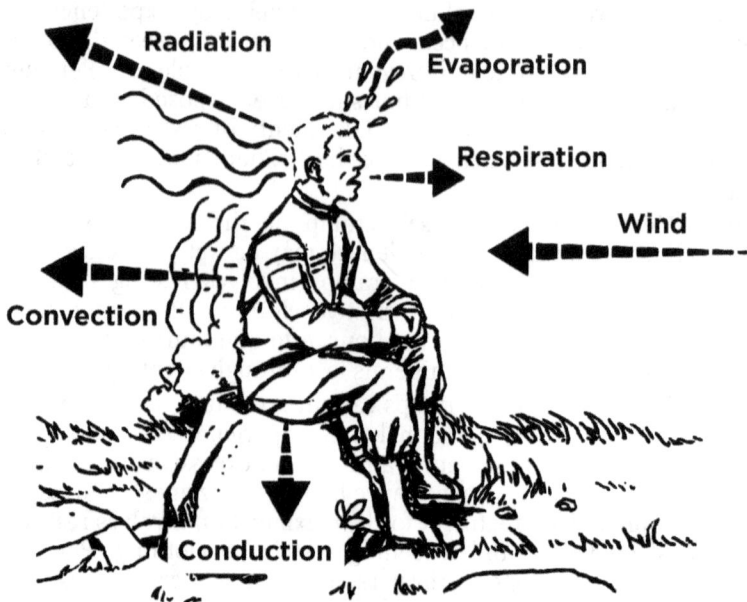

2-143. Radiation is the transfer of heat waves from the body to the environment or from the environment back to the body. It is the primary cause of heat loss. Radiation can contribute to heat loss of as 45 percent of body heat through exposed hands, feet, head, and neck, depending on the outside temperature. Use clothing to insulate exposed areas of the body. Convection is heat movement to or from an object by means of air or wind. The human body is always warming a thin layer of air next to the skin. When this warm layer of air is removed by convection, the body cools down.

2-144. A major function of clothing is to keep a warm layer of air close to the body. By removing or disturbing this warm air layer, wind can reduce body temperature. Conduction is the transfer of heat from one object to another. Extreme examples of rapid heat transfer include deep frostbite and third-degree burns. In hot conditions, sitting on a cool rock in the shade can conduct some excess heat away from the body. Evaporation is the process by which liquid changes into vapor. Heat within the liquid escapes to the environment. The body uses this method to regulate its core temperature (sweat) in hot, dry climates.

2-145. Clothing made of lightweight materials (cotton or linen) which have an open weave absorb sweat, allow air to circulate around the body, and help evaporation. Respiration works on the combined processes of convection, evaporation, and radiation. Air inhaled when breathing is rarely the same temperature as the air in lungs.

Chilblains and Frostbite

2-146. You can identify chilblains or frost nip by red, swollen, hot, tender, itching skin. Continued exposure to the cold could lead to infected skin lesions. The affected area usually responds well to locally applied warming (body heat). Do not rub or massage the area.

2-147. Frostbite, superficial and deep, are the next stages of cold injuries personnel should be aware of and treat if signs appear. Superficial frostbite will produce redness and blistering within twenty-four to thirty-six hours, and potential shedding of skin. Deep frostbite will be preceded by superficial frostbite, but the skin will be painless and pale yellowish in color. The spot will appear wooden, waxy, or solid to the touch and blisters may form within twenty four to thirty six hours.

2-148. To treat frostbite personnel should not attempt to thaw the affected areas by placing them too close to an open flame. The loss of feeling in the affected limb could lead to a severe destruction of tissue. Instead, thaw the affected area using water that is between 99 and 109° Fahrenheit (F) (37 and 43° Celsius (C)). Keep the tissue warm; decrease constricted clothing but increase exercise and insulation. Protect the affected area from further injury.

Immersion Foot

2-149. Immersion foot is caused by the foot being repeatedly exposed to wet, warm or cold conditions. A person's mobility can be affected in a severe case of immersion foot. The affected areas will be cold and numb. As the feet warm up, they may begin to feel hot with burning, shooting pains. In an advanced stage of immersion foot, the affected skin will be bluish and pale in color. The distal pulse will weaken, and blistering and swelling may occur. Casualties may also experience heat hemorrhaging and potential gangrene.

2-150. To treat immersion foot, gradually warm the affected area by exposing it to warm air. Do not massage or moisten the skin. Make sure to protect the affected area from trauma, dry the feet, and avoid walking. Personnel can prevent immersion foot by changing out of wet socks throughout the day. Washing and drying the feet daily also helps.

Hypothermia

2-151. Hypothermia is one of the most dangerous cold injuries personnel can experience. It can occur rapidly, and the temperatures do not have to be very cold to contract hypothermia. Isolated personnel are predisposed to hypothermia because of lack of food, potential dehydration, and little rest. Symptoms of hypothermia are coldness, wetness, and uncontrollable shivering. The body's core temperature will drop below 95° F (35° C).

2-152. The victim's movements can become uncoordinated, and unconsciousness may occur. Shock and coma may occur as the body temperature drops. To treat hypothermia, get victims out of wet clothing and into dry clothing, and have them slowly consume warm fluids. Begin to warm the body evenly and immediately. Keep victims dry and protect them from the elements. For severe hypothermia, quickly stabilize the victim's body temperature—this can include bare skin-on-skin contact inside a sleeping bag—and then prevent further heat loss.

Dehydration

2-153. The body continually uses water to support its processes. A person typically loses 1.5 liters of water a day through urine. Rapid fluid loss may also come from vomiting and diarrhea (severe diarrhea can increase fluid loss at up to 25 quarts in a 24 hour period). Another large source of water loss in our bodies is through sweating. Water is also lost from our body through breathing. Exposure to extreme hot and cold climate conditions and high altitudes expedite water loss. In extreme hot climates, the body can lose up to 3.5 quarts of water per hour when exposed to the elements. Additionally, intense activity burns, and illness, particularly the type that results in diarrhea, increases risk of dehydration.

2-154. To prevent dehydration, isolated persons should avoid or limit work during the hottest hours of the day. Sweat—not water—should be rationed when water is limited. If water is plentiful, drink enough water to ensure a steady urine output throughout the day. However, over-hydration can cause low serum sodium levels that can lead to death. Increase water intake in extreme climates. The body uses water more efficiently when small amounts of water are consumed at regular intervals. Isolated persons should always drink water when eating. Personnel acclimated to the environment require less water to maintain hydration.

2-155. Do not use alcoholic beverages, urine, blood, and seawater as water substitutes. Alcohol causes dehydration and impairs physical and mental abilities. Urine contains harmful toxins the body is attempting to excrete. It is also comprised of about 2 percent salt. Blood is salty and contains proteins, and therefore requires water to digest. It is considered a food and may transmit disease if it is consumed without cooking it first. Seawater is comprised of about 4 percent salt and results in dehydration.

2-156. Common signs and symptoms of dehydration include—

- Dark urine with strong odor.
- Low urine output.
- Dark, sunken eyes.
- Fatigue.
- Emotional instability.

- Skin inelasticity.
- Capillary refill delayed under fingernails.
- Trench line down center of tongue.
- Thirst. Thirst is not an adequate indicator of the need to drink water. Once personnel are thirsty, they are already dehydrated.

2-157. Signs and symptoms can be assessed to approximate the degree of dehydration personnel are experiencing:

- Symptoms can be described only by the person feeling them. If you are experiencing pain, no one knows unless you tell him or her. It is the same with dizziness, numbness, light-headedness, fatigue, vision disturbances, ringing in your ears, and a whole host of other feelings. Anyone who is not in your body is only going to know about these experiences if you describe them.
- That does not mean other people do not notice when you are not feeling well. If your face is pale, you are unstable when you walk, or you are sweating, then you are showing signs. Signs have to be seen and read by someone (e.g., another Soldier or healthcare provider) rather than felt. Signs are just what they sound like: indicators of a problem.
- A 2% loss results in thirst.
- A 10% loss results in dizziness, headache, inability to walk, and a tingling sensation in the limbs.
- A 15% loss results in dim vision, painful urination, swollen tongue, deafness, and a numb feeling in the skin.
- Death can result from a loss of greater than 15%.
- Carry Alka-Seltzer tablets to add carbonation to water to help speed its use in the large intestine.

INGESTION POISONING

2-158. Poisons can come from misidentified plants, animals, toxic water sources, toxic fungi, and chemicals. Poisons can enter the body through four means: eyes, nose, mouth and skin contact. The treatments for poison ingestion under field conditions include inducing vomiting and taking about 1 gram (0.04 ounce) of activated charcoal. Activated charcoal will attach and absorb the poison and carry it out of the gastrointestinal system. Activated charcoal can be made from a wide range of carbon rich materials like wood, coal, bones, coconut shells, nut shells and peat. The first step in making activated charcoal is to take the material and bring it to a temperature between 600-900 degrees Fahrenheit in a survival fire. Burn the material for about 4 ½ hours. Cool the burnt material, crush it into a fine powder, and use for survival medicine remedies and water filtration.

2-159. In some cases, a laxative may be required to help the activated charcoal work to its full potential. If the poison is an acid, an alkali, or a petroleum product, vomiting should be avoided because the substance can burn the esophagus. These types of products can also be inhaled into the lungs during vomiting, causing pneumonia. Aggressively hydrate the body with clean, treated potable water to prevent renal failure and dilute the poison in the victim's system.

PERSONAL HYGIENE AND SANITATION

2-160. Application of the following simple guidelines regarding personal health and hygiene will enable the isolated person to safeguard personal health and the health of others while detained or captive.

2-161. Stay clean (daily regimen) by—

- Minimizing infection by washing (use white ashes, sand, or loamy soil as soap substitutes).
- Combing and cleaning debris from hair.
- Cleansing mouth and brushing teeth.
  - Use hardwood twig as toothbrush (fray it by chewing on one end then use as brush).
  - Use single strand of an inner core string from parachute cord for dental floss.
  - Use clean finger to stimulate gum tissues by rubbing.
  - Gargle with saltwater to help prevent sore throat and aid in cleaning teeth and gums.
- Cleaning and protecting feet.
  - Change and wash socks
  - Wash, dry, and massage.
  - Check frequently for blisters and red areas.
  - Use adhesive tape/mole skin to prevent damage.
- Exercising daily.
- Preventing and controlling parasites.
  - Check body for lice, fleas, ticks.
  - Check body regularly.
  - Pick off insects and eggs (DO NOT crush).
  - Wash clothing and use repellents.
  - Use smoke to fumigate clothing and equipment.

AVOIDING ILLNESS

2-162. You can avoid many different kinds of illnesses and infections by practicing good sanitation and hygiene. Maintaining a clean body and living area will prevent the spread of germs and bacteria whether you are alone or in a group. It will also you to stay organized and protect items from animal contact.

2-163. Rules for avoiding illness:

- Purify all water obtained from natural sources by using iodine tablets, bleach, or boiling for 5 minutes.
- Locate latrines 200 feet from water and away from shelter.
- Wash hands before preparing food or water.
- Clean all eating utensils after each meal.
- Prevent insect bites by using repellent, netting, and clothing.
- Eat varied diet.
- Try to get 7-8 hours sleep per day.

Note: Purify all water obtained from natural sources before consumption, if possible.

2-164. Application of the following simple guidelines regarding personal hygiene will enable you to safeguard personal health and the health of others in a survival situation:

- Do not soil the ground in the camp area with urine or feces; use latrines. When no latrines are available, you should dig "cat holes" and cover your waste with earth or sand.
- Never put fingers and other contaminated objects into the mouth. Wash hands before handling any food or drinking water, before using the fingers in the care of the mouth and teeth, before and after caring for the sick and injured, and after handling any material likely to carry disease or germs.
- Clean and disinfect all eating utensils in boiling water after each meal.
- Thoroughly clean the mouth and teeth at least once each day.
- Avoid insect bites by keeping the body clean; wearing proper protective clothing; and using a head net, improvised bed nets, and insect repellents.
- Exchange wet clothing for dry clothing as soon as possible to avoid unnecessary body heat loss. Keep clothing clean and in good repair.
- Do not share personal items such as canteens, towels, toothbrushes, handkerchiefs, and shaving items.
- Aim for seven or eight hours of sleep each night.
- Keep hair clean because dirty hair can be a haven for bacteria, fleas, lice, and other parasites.
- Take care of feet; clean them and keep them dry.

INTESTINAL PARASITES

2-165. Though seldom fatal, worms are significant as they can lower an individual's general resistance to other illnesses. Common identifiers of worms include severe rectal itching, insomnia, and restlessness. Prevention can be a simple and readily obtainable goal. Undercooked or contaminated food, poor sanitation, and infected water sources are main causes of intestinal parasites and worms.

2-166. Eat hot peppers as treatment as they contain certain substances chemically similar to morphine. They are effective as a counterirritant for decreasing bowel activity. Grind pumpkin seeds and mix with water creating a "medicinal" porridge or make a cup of tea made from one to two grams (about ½ to 1 ounce) of thyme and drink several times a day.

2-167. Honey makes an ideal topical wound ointment to reduce infection and promote healing. Its hypertonicity[1], low pH, and inhibins[2] provide antimicrobial activity against many types of bacteria and at least two types of fungi.

---

1. "Hypertonicity" refers to honey's relatively high osmotic pressure in relation to bacterial cells; this causes water to be drawn out of the bacterium, killing it.
2. Inhibins: antimicrobial substances, several of which are present in honey, including hydrogen peroxide (produced by the breakdown of glucose oxidase in the presence of free water).

CHAPTER 3

WATER

This chapter discusses the criticality of water to the isolated person. It includes useful information concerning hydration, water sources and indicators, as well as ways of procuring and preparing potable water.

HYDRATION CONSIDERATIONS

3-1. During isolation, the isolated person is faced with a constant dilemma — what actions must be done and how to perform them with the least expenditure of effort or capability while maintaining personal security. Water is one of the isolated person's most urgent needs and the procurement of potable water is an essential task that must be performed in all operational environments.

3-2. There are several general rules that apply to water and hydration. These include—

- If you have to choose between running water and stagnant water, choose running water but make sure it is also appropriately treated.
- Do not eat snow. The body has to heat the water and melt the snow once you eat it and is a foolish expenditure of calories that are already in short supply.
- Do not drink saltwater, even a small amount.
- Do not drink water found in natural depressions. It is stagnant water and is often heavily contaminated with pesticides, garbage and other debris.
- Do not drink urine or alcoholic beverages.
- Consider all surface water to be contaminated with human and animal waste, with the possible exception of very high mountain streams or springs found in uninhabited areas.

3-3. At a temperature of 68° F with limited physical activity, personnel will normally require 2 to 3 quarts of water a day to maintain efficiency. Water is necessary to replace what is lost through daily functions. Personnel lose about 1.4 quarts of water through urine loss, 1.0 quart of water through sweat, and 0.2 quarts of water through fecal matter per day. Water loss increases with heat exposure.

3-4. When personnel are exposed to high temperatures, water loss from sweat increases to as much as 3.5 quarts per hour. At this rate, body fluids are quickly depleted. Physical activity increases loss of water in two ways: a) increased respiration and b) increased sweating due to excessive body heat. The time and effort required to obtain water and the decrease in the thirst response in cold weather favors the development of dehydration. An isolated person who is burned may lose up to 5 quarts of water per day. When ill, they will lose water through vomiting and diarrhea.

WATER SOURCES AND INDICATORS

3-5. Almost any environment has water present to some degree. Table 3-1 lists possible sources of water in various environments. It also provides information on how to procure the water as well as how to make it potable.

PROCURING WATER

3-6. You can procure water using a variety of different methods, depending upon the source. The first step to procuring water is obtaining a container that can temporarily hold water to enable purification and that can be sealed for long-term storage. If issued containers are unavailable or damaged, isolated persons can improvise them from non-porous materials. Water containers can be constructed out of plastic bags, condoms (placed inside a sock to protect the bladder), bamboo segments, life preserver units, birch bark and pitch containers; as well as gourds, tarps, sheets of plastic and waterproof evasion charts (see also "Water Containers" on page 274).

Table 3-1. Means to Make Potable Water

Frigid Areas	
Sources of water	• Snow and ice

Means of obtaining and/or making potable	• Melt and purify
Remarks	• Do not eat without melting! Eating snow or ice can reduce body temperature and lead to more dehydration. • Snow or ice are no purer than the water from which they come. • Sea ice that is gray in color or opaque is salty. Do not use it without desalting it. Sea ice that is crystalline with a bluish cast has little salt in it.

At Sea

Sources of water	• Sea • Rain • Sea ice
Means of obtaining and/or making potable	• Sea: use desalinator. • Rain: catch rain in tarps or in other water-holding containers.
Remarks	• Sea: do not drink seawater without desalting. • Rain: if tarp or water-holding material is coated with salt, wash it in the sea before using (very little salt will remain on it). • Sea ice: see previous remarks on frigid areas.

Beach

Sources of water	• Ground • Fresh

Means of obtaining and/or making potable	• Dig a hole deep enough to allow water to seep in. Obtain rocks, build fire, and heat rocks; drop hot rocks in water; hold cloth over hole to absorb steam; wring water from cloth. • Dig behind first group of sand dunes. This will allow the collection of fresh water.
Remarks	• Alternate method if a container or bark pot is available: Fill container or pot with seawater; build fire and boil water to produce steam; hold cloth over container to absorb steam; wring water from cloth.

Desert

Sources of water	• Ground • In valleys and low areas • At foot of concave banks of dry rivers • At foot of cliffs or rock outcrops • At first depression behind first sand dune of dry lakes • Wherever you find damp surface sand • Wherever you find green vegetation
Means of obtaining and/or making potable	• Dig holes deep enough to allow water to seep in.
Remarks	• In a sand dune belt, any available water will be found beneath the original valley floor at the edge of dunes.

3-8. Water containers such as jerry cans in a cold weather environment should be stored upside down to enable their use when frozen. Since water freezes from top to bottom, any unfrozen water would be on the bottom where it is accessible. If the container is stored improperly ice will block the opening making the water inaccessible.

3-9. Isolated persons should always take into consideration the safety of different sources and procure water from the safest source. Some sources such as plants and certain types of precipitation do not require purification and make a great choice for personnel who do not have the ability to purify water. Other sources can provide large quantities of water on demand, but require purification. Avoid water that has a foul odor or smell, and water potentially contaminated with toxins. Faster moving water contains more oxygen and typically fewer harmful microorganisms than stagnant water and is usually a better option to procure water.

RAIN, SNOW, AND ICE

3-10. Fresh rainwater collected into containers does not require purification. However, rain that runs down leaves and other vegetation on its way to the container can be contaminated and should be purified. If you allow rain to wash contaminants off overhead leaves for at least 15 minutes, it is then safe to collect and drink immediately.

3-11. Collect rainwater by putting out improvised nonporous items to catch the rain as it falls from the sky. Isolated persons can also construct an elaborate device that will catch a large volume of rain while they perform other tasks. An example would be the bamboo rain catcher (see Figure 3-1). Construct this apparatus by making a set of framework off the ground out of bamboo. Split the bamboo in half and layered in a shingle-type formation in a flat surface formation. Place one end higher than the other with a collection channel placed at the bottom that channels the water to a collection container.

Figure 3-1. Bamboo Water Catch (or Rain Catcher)

3-12. Use fresh, clean snow as a water source. Place containers filled with snow near a fire or between clothing layers to allow body heat to melt it. Adding water to snow will help it melt faster. Avoid directly eating snow as it lowers the body's core temperature and requires the body to use more energy reserves than if the snow is heated externally.

3-13. Isolated persons can construct a water machine that will produce water from snow while they are completing other tasks. Place snow on any porous material (such as cotton or parachute nylon), gathering up the edges, and suspending the "bag" of snow from any support structure near a fire. Radiant heat will melt the snow, and water will drip to the lowest part of the bag. Place a water container under the snow machine to collect the water. You need only to fill the machine with snow as required and empty the container of water when full (see Figure 3-2). Remember though, snow and ice are no purer than the water from which they come.

Figure 3-2. Water Machine

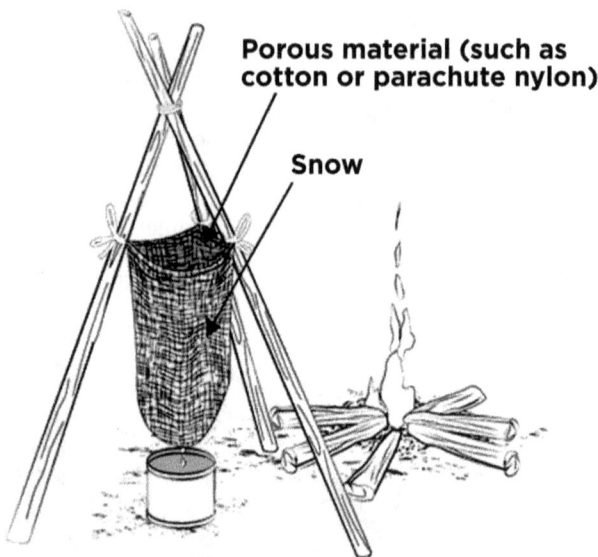

Porous material (such as cotton or parachute nylon)

Snow

3-14. Sea ice provides a good source of water. You can use old sea ice that is crystalline and deep bluish or black in color and has rounded edges when shattered. This type of old sea ice has the least amount of salt content and is drinkable. Avoid gray-colored or opaque sea ice due to its high salt content. New sea ice has sharp edges when shattered. Do not consume until it is desalted. Ice provides a more concentrated source of water than snow and requires less heat to do so. Melt snow or ice with minimal effort by placing it on a dark piece of material in sunlight.

DEW

3-15. Collect dew by tying fine grasses, a sponge, or cloth around the lower legs and ankles and walking through tall grass to soak up the dew water. Collect up to one quart of water or more by wringing it out into a container.

WATER FROM PLANTS

3-16. If no open water source is found, there are plants in many areas around the globe that will produce drinking water. Caution must be taken to prevent obtaining water from a poisonous source. Positive plant identification is the safest and easiest method for determining whether water inside the plant is potable. Water procured from the inside of a plant when sealed off from outside contaminants does not need to be purified, but the means are available to do so it is recommended as a precaution.

Green Bamboo

3-17. Green bamboo stores water inside its individual segments. Shake the plant to listen for water sloshing inside the plant. If it contains water, the bamboo is broken open to obtain it. The top of the green bamboo can be cut and the end bent and tied over toward the ground with a water container under the cut top to let the water drain out for collection (see Figure 3-3).

Figure 3-3. Procuring Water from Green Bamboo

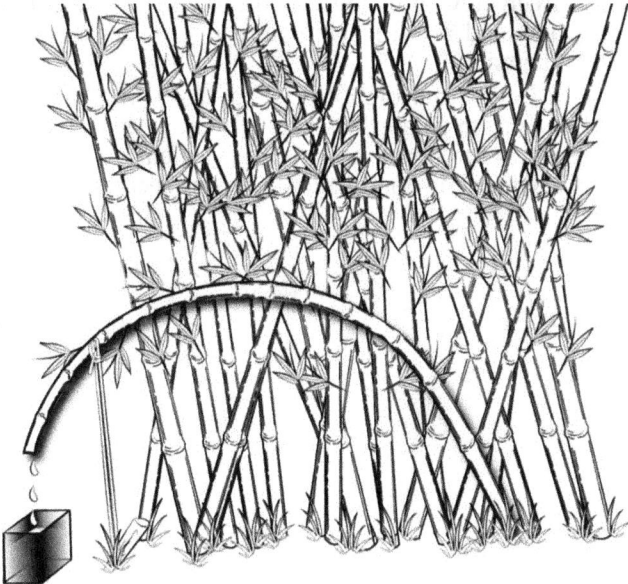

Water Vines

3-18. Many different types of vines contain drinkable water. Vines can be 50 feet to several hundred feet in length and 1 to 6 inches in diameter. They grow from the ground at the base of a tree and climb toward the canopy to obtain sunlight. Water vines are usually soft and easily cut. Small vines may be twisted or bent easily and are usually heavy because of the water content. The water from these vines should be tested to ensure it is safe unless positive identification of a safe plant is made. Use these steps to test the water from these vines:

Step 1 Notch the vine and watch for sap running from the cut. If milky sap is observed, the vine should be discarded. If not, move to step 2.

Step 2 Cut out a section of the vine and hold it vertically. Observe the liquid as it flows out. If the liquid is cloudy or milky-colored, the vine should be discarded. If not, move to step 3.

Step 3 Let some of the liquid flow into the palm of the hand and observe it. If the liquid changes color it should be discarded. If not, it may be tasted. Any liquid that has a woody or sweet taste is safe for drinking. Vines that produce a liquid with a sour or bitter taste are avoided.

3-19. Once the water is determined safe for drinking, it can be procured in larger quantities. The first cut is made high on the vine above the ground. The second cut is made low near the ground. This method maximizes the amount of water procured per vine and allows the water to freely flow through the plant. When drinking water from the vine, you should avoid placing your lips on the bark as it may contain skin irritants.

3-20. Obtain water from the rattan palm and spiny bamboo in the same manner as from vines. (See Figure 3-4 on page 93). The slender stem or runner of the rattan palm is an excellent water source. The joints are overlapping in appearance, as if one section is fitted inside the next.

Figure 3-4. Vine Water Procurement Technique

Banana Plants

3-21. Water can be obtained from banana plants by cutting a banana plant down into a long section which can be easily handled. Take the section apart by slitting from one end to the other and pulling off the layers one at a time. A strip 3 inches wide, the length of the section, and just deep enough to expose the cells should be removed from the convex side. Fold this section toward the convex side to force the water from the cells of the plant. The layer must be squeezed gently to avoid forcing out any tannin (an astringent which has the same effect as alum) into the water.

3-22. Another technique for obtaining water from the banana plant is by making a banana-well. The tree is cut down leaving a 12 inch (30 centimeter) stump. A bowl is hollowed out of the plant stump, fairly close to the ground, by cutting out and removing the inner section.

(see Figure 3-5). Water will be drawn into the bowl by the plant's roots, the first of which contains a concentration of tannin that is bitter and should not be consumed. Bitter water is discarded until it is palatable, about 3 fillings. A leaf from the banana or other plant should be placed over the bowl while it is not being used in order to prevent contamination. The roots provide refill the bowl for approximately 4 days. This same method can be used to procure water from plantain trees and sugarcane.

Figure 3-5. Water from Banana Stump

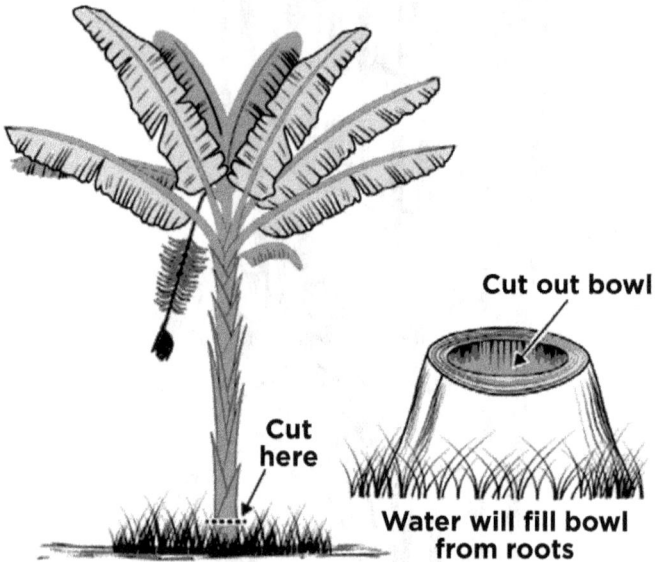

Cut out bowl

Cut here

Water will fill bowl from roots

Water Trees

3-23. Water trees in the tropics can be a valuable source of water. It is important to properly identify the tree as the sap of some trees can be very dangerous. Water trees can be identified by their blotched bark which is fairly thin and smooth. The leaves are large, leathery, fuzzy, and evergreen, and may grow as large as 8 or 9 inches in length. The trunks may have short outgrowths with fig-like fruit on them or long tendrils with round fruit comprised of corn kernel-shaped nuggets.

3-24. The tree can be tapped in the same manner as a rubber tree, with either a diagonal slash or a "V." When the bark is cut into, it will exude a white sap, which, if ingested, causes temporary irritation of the urinary tract. This sap dries up quite rapidly and can easily be removed. The cut should be continued into the tree with a spigot (bamboo, knife, or such) at the bottom of the tap ("V") to direct the water into a container. The water flows from the leaves back into the roots after sundown, so water can be procured from this source only after sundown or on overcast (cloudy) days.

Coconut Water

3-25. Mature coconuts contain an oil that can cause diarrhea when consumed in excess. There is little concern if used in moderation or with a meal, but the symptoms may be present if consumed on an empty stomach. Green, unripe coconuts about the size of a grapefruit are the best for use because the fluid can be taken in large quantities without negative side effects. There are more fluids and less oil in unripe coconuts, so there is less possibility of diarrhea. Figure 3-6 on page 96 depicts how you can locate, obtain, and open a coconut for nutrients. The husk is first removed to expose the nut. The nut is then cracked or notched open to access the liquid inside.

Figure 3-6. How to Open a Coconut

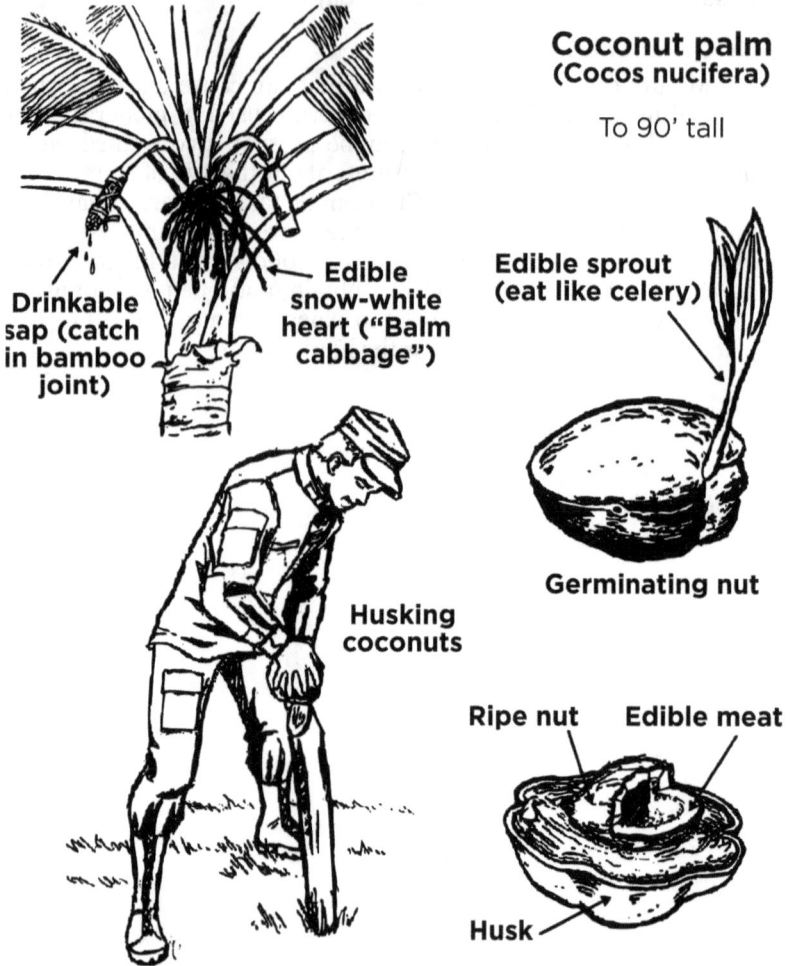

Coconut palm
(Cocos nucifera)

To 90' tall

Drinkable
sap (catch
in bamboo
joint)

Edible
snow-white
heart ("Balm
cabbage")

Edible sprout
(eat like celery)

Germinating nut

Husking
coconuts

Ripe nut   Edible meat

Husk

Vegetation Bag Still

3-26. Water can also be collected through the use of a vegetation bag still (see Figure 3-7 on page 97). This method requires a plastic bag filled one-half to three-fourths full with non-poisonous vegetation. A rock is then placed in the bag. Before closing the bag, it is filled with air by turning the opening into the wind. It is then closed air-tight, around the tubing if used, and placed in the hot sun, optimally located on a sun-facing slope. The sunlight heats the vegetation and distills the water vapor out of the vegetation to collect onto the surface of the bag and down to a collection point that has been modified by creating a low spot with a rock. Surgical tubing or a reed is used like a straw to drink directly from the bag. All condensed water is collected before the sun goes down. When dusk arrives, the water draws back into the vegetation. The vegetation bag is changed after extracting most of the water to ensure maximum output.

Figure 3-7. Vegetation Bag Still

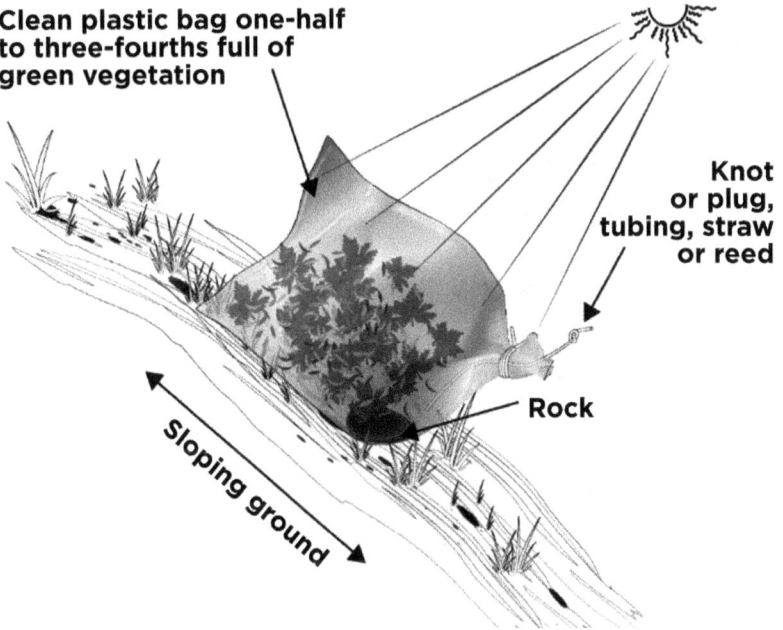

Clean plastic bag one-half to three-fourths full of green vegetation

Knot or plug, tubing, straw or reed

Rock

Sloping ground

Transpiration Bag Still

3-27. A water transpiration bag still is very similar to the vegetation bag still, only easier (see Figure 3-8 on page 98). Tie a plastic bag around the leafy branch of a tree or plant. The process requires an airtight seal and that the branch be weighted to make a collection point for the water. All water is collected before sundown so the plant does not reabsorb the water into its system. The same limb can be used for 3~5 days without causing long-term harm to the limb: it heals itself within hours of removing the bag.

Figure 3-8. Transpiration Bag Still

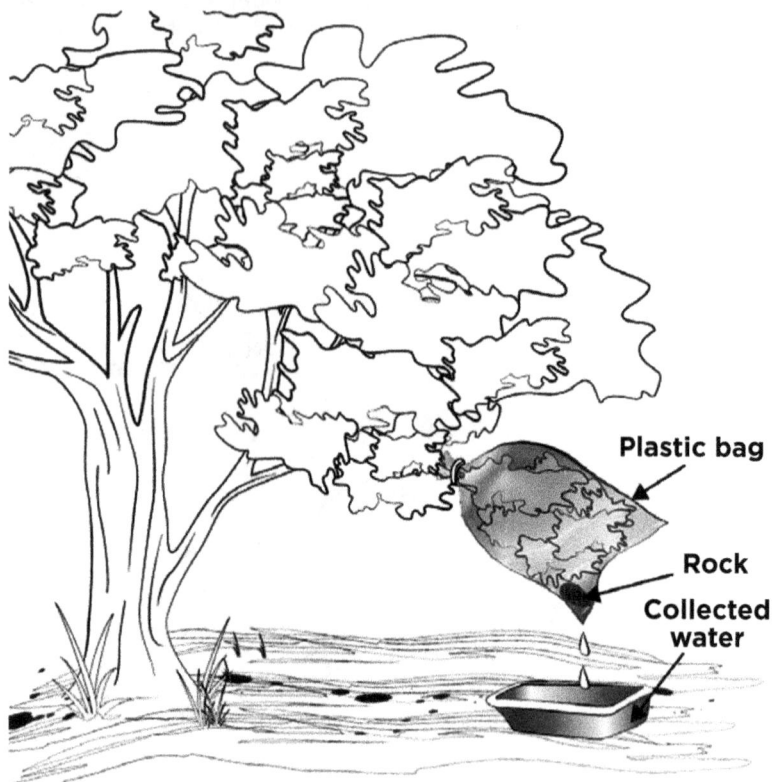

BELOW-GROUND SOLAR STILL

3-28. To make a below-ground solar still, you will need a digging tool, a container, a clear plastic sheet, and a rock (see Figure 3-9 on page 99). A tube can be used as a straw to drink the water without disturbing the still. When the still is opened, the moist, warm air that has been accumulated is released and the distillation process is slowed.

3-29. A site is selected where you believe the soil contains moisture (such as a dry streambed or a low spot where rainwater has collected) to pull water from the soil. The soil at this site should be easy to dig and sunlight must hit the site most of the day. Plants can also be cut and placed along the wall of the still as the water source. Additionally, the below-ground still can be used to separate drinkable water from saltwater or polluted water.

Figure 3-9. Below-ground Solar Still

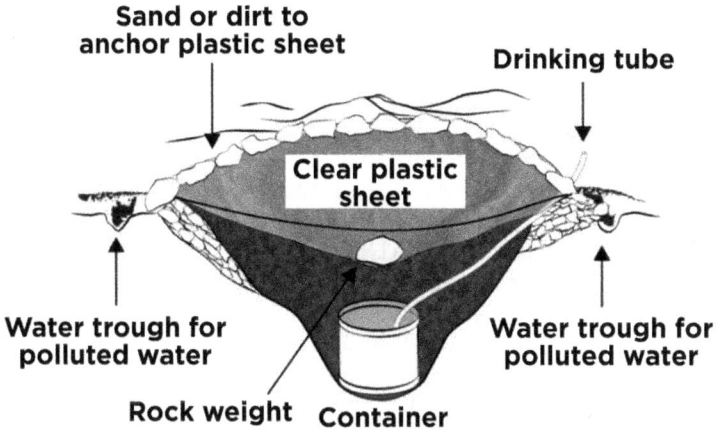

Sand or dirt to anchor plastic sheet

Drinking tube

Clear plastic sheet

Water trough for polluted water

Water trough for polluted water

Rock weight **Container**

3-30. Use the following steps to construct a below-ground solar still:

Step 1 Dig a bowl-shaped hole about 3 feet (1 meter) across and 24 inches (60 centimeters) deep.

Step 2 Dig a sump in the center of the hole. The sump's depth and perimeter will depend on the size of the container that to be placed in it. The bottom of the sump should allow the container to stand upright.

Step 3 Anchor the tubing to the container's bottom by forming a loose overhand knot in the tubing.

Step 4 Place the container upright in the sump.

Step 5 Extend the unanchored end of the tubing up, over, and beyond the lip of the hole. Place the plastic sheet over the hole, covering its edges with soil to hold it in place.

Step 6 Place a rock in the center of the plastic sheet.

Step 7 Lower the plastic sheet into the hole until it is about 16 inches below ground level. It now forms an inverted cone with the rock at its apex. Ensure that the cone's apex is directly over the container. Also, ensure that the plastic cone does not touch the sides of the hole because the earth will absorb the condensed water.

Step 8 Put more soil on the edges of the plastic to hold it securely in place and to prevent the loss of moisture.

Step 9 Plug the tube when not in use to keep the moisture from evaporating and to keep insects out.

3-31. If saltwater or polluted water is used as the moisture source, dig a small trough outside the hole about 10 inches deep and 3 inches wide. Pour the water in the trough. Be sure not to spill any polluted water around the rim of the hole where the plastic sheet touches the soil. The trough holds the water and the soil filters it as the still draws it. The water then condenses on the plastic and drains into the container. The process works extremely well for polluted and saltwater. Three or more of these devices should be constructed to produce enough water for one person's needs.

3-32. Isolated persons should attempt to control energy output when building these devices to control body water loss that would defeat the purpose of the device. Use caution as this process is time- and energy-intensive and may not be worth the return of water quantity.

BEACH WELLS AND SALTWATER

3-33. Saltwater must be desalinated before drinking. This can be done by using a desalination pump or solar still (see Paragraph 3-28 onward). When on a saltwater beach isolated persons primarily search for freshwater. Freshwater can be procured through the use of a beach well, a hole behind the first group of sand dunes dug deep enough to hit the water table (see Figure 3-10). Wood is placed around the sides of the hole to stabilize it and control cave-ins from the sand. This water should be fresh and salt-free. Alternately, isolated persons can dig a hole on the shoreline and let it fill with saltwater. Superheated rocks are then placed into the water resulting in steam, which is collected with a rag and transferred to a holding container, salt free and ready to drink.

Figure 3-10. Beach Well

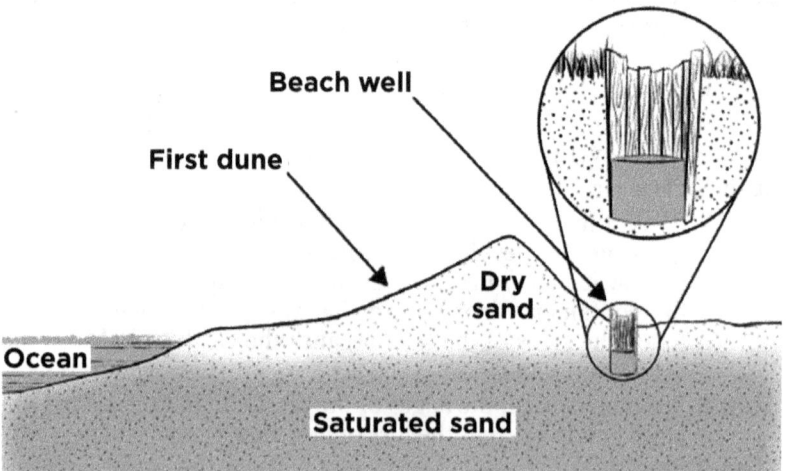

SEEPAGE WELLS AND SUB-SURFACE WATER

3-34. A seepage well is a hole dug into the ground that will tap into the water pooled or moving below the ground. If located, this water will pool in a seepage well and can be collected. (See Figure 3-11). This technique can be used in most environments. Outside bends of a dry riverbed, the foot of cliffs or rocks, low areas protected from the sun, or damp subsurface sand would be ideal locations for a seepage well.

Figure 3-11. Sub-surface Water

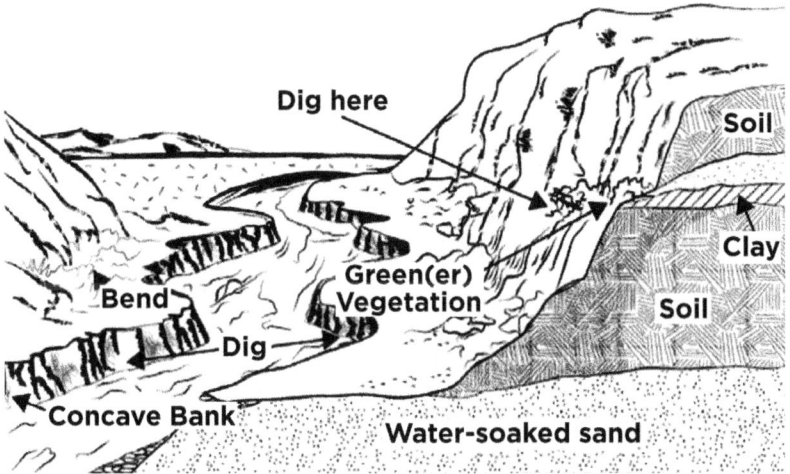

PREPARING WATER

3-35. Isolated persons should remain hydrated with good potable drinking water, making sure to purify any water that may be contaminated. Water that is high in sediment or debris content can be filtered to increase palatability and consumption. Rainwater and water collected from plants is usually safe for drinking; however, water from open sources such as lakes, streams, ponds, and swamps—especially any water near human settlements or in the tropics—will need to be purified. Springs are typically the safest open source to drink from, if the seepage is located, but should be purified whenever the means are available as a precaution. When possible, all water should be purified by boiling, chemical disinfectants, or commercial filters that purify water through the filtration of extremely fine particulates.

DRINKING WATER HAZARDS

3-36. Persons drinking non-potable water may become ill from ingesting harmful organisms that can cause diarrhea and lead to death. Two prevalent pathogens found in most water sources throughout the world are—

- **Giardia**. Giardia causes giardiasis. This is an illness caused by Giardia lamblia, a single-celled, microscopic parasite that lives in the intestines of people and animals. The parasite is passed in the bowel movement of an infected person or animal. It is characterized by an extreme case of runny diarrhea accompanied by severe cramps, weight loss, and dehydration. The symptoms usually appear one to two weeks after the infection with the parasite. In a healthy person, the symptoms can last from four to six weeks. If giardiasis is suspected, isolated persons should increase their intake of potable water to remain hydrated.

- **Cryptosporidium**. This also a small parasite, which causes cryptosporidiosis. It is much like giardiasis, only more severe and prolonged, and there is no known cure but time. Diarrhea may be mild and can last from 3 days to several weeks.

Note: The only effective means of neutralizing cryptosporidium is by boiling or by using a commercial microfilter or reverse-osmosis filtration system. Chemical disinfectants such as iodine tablets or bleach have not shown to be 100 percent effective in eliminating cryptosporidium.

3-37. Examples of other water-related illnesses and organisms include:

- **Dysentery**. Personnel may experience severe, prolonged diarrhea with bloody stools, fever, and weakness.
- **Cholera** and **typhoid**. Personnel may be susceptible to these diseases regardless of inoculations. Cholera can cause profuse, watery diarrhea, vomiting, and leg cramps. Typhoid symptoms include fever, headache, and loss of appetite, constipation, and bleeding in the bowel.
- **Hepatitis A**. Symptoms include diarrhea, abdominal pain, jaundice, and dark urine. This infection can spread through close person-to-person contact or of ingestion contaminated water or food.
- **Flukes**. Stagnant, polluted water-especially in tropical areas-often contains blood flukes. If swallowed, flukes bore into the bloodstream, live as parasites, and cause disease.
- **Leeches**. If leech is swallowed, it can hook onto the throat passage or inside the nose and create a wound before moving to another area. Each wound can become infected.

FILTRATION

3-38. If the water is muddy, stagnant, and foul-smelling, it can be cleaned by placing it in a container and letting it stand for several hours so the sediment settles to the bottom. The water can be aerated by pouring it between two containers: this will improve the taste. If time and materials are available, a water filter can be improvised by using material or a large container and building layers of material in the container. The filter system is made from natural materials like sand, crushed rock or gravel, and charcoal. (see Figure 3-12). The filter will remove large and small particles from the water, aerate it for taste, and allow you to filter larger amounts of water for consumption.

Figure 3-12. Water Filtering Systems

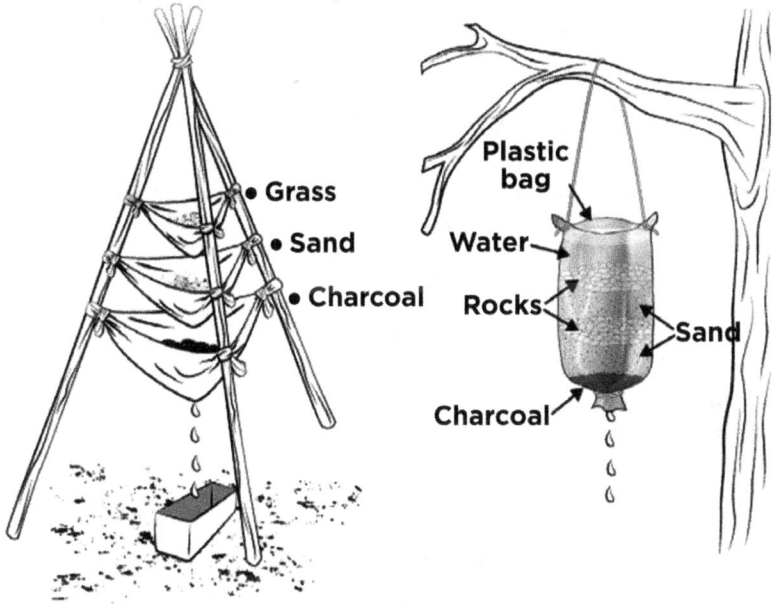

3-39. A sediment hole or seepage basin can be built by digging a hole alongside a water source allowing the ground to filter the water through the soil. The water will flow through the surrounding soil and pool in the hole, ready for purification. Charcoal from a fire can also be used to improve water. When added to water, and allowed to set for 45 minutes, it helps remove odor and absorbs agricultural and industrial chemicals. After water has been filtered to improve palatability, it still requires purification prior to consumption.

PURIFICATION

3-40. Rainwater and water collected from plants is usually safe for drinking; however, water from lakes, streams, ponds, swamps, and springs—especially any water near human settlements or in the tropics—will need to be purified. When possible, all procured water should be purified by boiling, sunlight, or chemical disinfectants (including purification filters: water filters that filter out extremely fine particulates and at the same time treat the water with chemicals).

BOILING METHOD

3-41. Boiling is the safest method for purifying water. Bringing water to a rolling boil for one minute at sea level will kill all waterborne pathogens. If above 8,000 feet, the boiling time should be extended to 3 minutes.

Note: These boiling times are different from the times required for sterilizing water for medical use.

CHEMICAL PURIFICATION

3-42. Chemicals can be used to purify water when necessary. When using chemicals, you must remember to let the water stand for the prescribed amount of time and then open the cap and let the purified water rinse any unpurified water from the threads of the container before touching it to your mouth.

Types of Chemical Purification

3-43. The following are types of chemical purification:

- **Purification tablets**. When using water purification tablets always follow the manufacturer's instructions for the recommended use of the product. Always verify that the tablets are within their specified expiration date to ensure proper dosage. These tasks should be completed before the mission.
- **Tincture of iodine**. Use 5 drops of 2 percent tincture of iodine per full quart of water. If the water is cloudy or cold use 10 drops to ensure proper purification. After the iodine is mixed in, allow the water to stand for 30 minutes before drinking. If the water is cold and cloudy, wait 60 minutes.
- **Povidone iodine**. Use 2 drops of 10 percent military-grade povidone iodine per full quart of water. Titrated povidone iodine is the civilian equivalent at only 2 percent strength, and so requires 10 drops per quart. Let the water stand for 30 minutes. If the water is cold and clear, wait 60 minutes. If it is very cold or cloudy, add 4 drops and wait 60 minutes.

WARNING: If you are allergic to shellfish, you may also be allergic to iodine. An alternative is the use of bleach or boiling.

- **Chlorine bleach**. Place 2 drops of chlorine bleach (5.25 percent sodium hypochlorite) for 1 quart of water and let it rest for 30 minutes. If cold and cloudy, wait 60 minutes. Remember that not all bleach is the same around the world; check the available level of sodium hypochlorite.

3-44. There are several options for purifying water available on the open market and not covered in this publication. If other options are used, read and follow manufacture's recommended instructions. When adding drink mix to the water to improve taste, always ensure that the water has completed the purification process before adding the mixture to the water.

Commercial Filters

3-45. Not all commercial filters achieve the same level of filtration. Filters that remove the smallest harmful microorganisms found in the area of operations should be selected. Mechanical water filters are simple to use and require no holding time. Filters do not add any poor taste and, in most cases, improve the taste of water. A water filter's effectiveness is not dependent on water temperature. Disadvantages to water filters and similar filters are that they add bulk and weight to equipment, and most filters do not remove viruses. Filters are expensive compared to chemical treatments. As a filter becomes dirty it takes more pressure to push the water through, and small cracks and channeling in the filter can let small microorganisms through the filter. Eventually filters will need cleaning or some sort of field maintenance.

3-46. Smallest-pore nano filters require higher pressure, as in reverse-osmosis filters. Freezing water within the filter element will compromise or destroy some filters. These types of filters provide little holding of contaminants and will clog rapidly; however, they can be easily stripped and cleaned by brushing and washing them without destroying the filter.

3-47. Maze or depth filters depend on a long, irregular labyrinth to trap organisms. Contaminants adhere to the walls of the filters or are trapped in the numerous dead-end tunnels. Granular media such as sand or charcoal, diatomaceous earth[1], or ceramic filters function as maze filters. A depth filter has a larger holding capacity for particles and lasts longer before clogging. These types of filters may be difficult to clean because particles can be trapped deep within them. Flow can be partially restored to a clogged filter by back-flushing or surface cleaning, which removes the larger particles trapped near the surface.

---

1. Diatomaceous earth is a soft sedimentary rock that can be crumbled into a fine powder for use in filtration.

CHAPTER 4

FOOD

This chapter discusses one of the isolated person's most urgent requirements. The isolated person must remember that the three essentials of survival—water, food, and shelter—are prioritized according to their estimate of the situation. We can live for weeks without food but it may take days or weeks to determine what is safe to eat and to trap animals in the area. Therefore, you need to begin food gathering in the earliest stages of isolation.

FOOD CONSIDERATIONS

4-1. Although you can live several weeks without food, you need an adequate amount to stay healthy. Without food, your mental and physical capabilities will deteriorate rapidly and you will become weak. Food provides energy and replenishes the substances that your body burns. Food provides vitamins, minerals, salts, and other elements essential to good health. Possibly more important, it helps morale.

4-2. The three basic sources of food are plants, animals (including fish and sea life), and commercial rations. In varying degrees, all three provide the calories, carbohydrates, fats, and proteins needed for normal daily body functions. Commercial rations should be saved to augment plant and animal foods, which will extend and help maintain a balanced diet.

4-3. The average person needs 2,000 calories per day to function at a minimum level. An adequate amount of carbohydrates, fats, and proteins without an adequate caloric intake will lead to starvation and cannibalism of the body's own tissue for energy. Signs of malnutrition/starvation include the following—

- Loss of body fat.
- Difficulty breathing.
- Lower body temperature.
- Reduced muscle mass.

- Fatigue.
- Longer healing time for wounds and illness.

NUTRITION

4-4. From the moment a piece of food enters the mouth, each bite is broken down and used by the body. Metabolism is the series of chemical reactions that transform food into components that can be used for the body's basic processes. Carbohydrates, fats, proteins, minerals, and vitamins are vital to proper function of the body.

Carbohydrates

4-5. Carbohydrates provide an immediate source of energy, or are stored for future use. When stored as glycogen in the liver and muscles they remain an effective and efficient energy source.

4-6. Excess calories from carbohydrates (just like excess protein and fat) convert to fat and serve as an energy source for future activity. Because of their role as a primary fuel source for physical and cognitive activity 45 percent to 65 percent of your calories should come from carbohydrates. Good sources of carbohydrates include fruits, vegetables, grains, and dairy products.

Fats

4-7. Fats are over twice as calorically dense as carbohydrates and protein and can provide an important source of energy. Fats come in many forms and are essential for normal body function. In addition to being a good energy source, fats play a key regulatory role in numerous biological functions and serve as a carrier for the absorption of the fat-soluble vitamins A, D, E and K and carotenoids[1]. The recommended total fat intake is between 20 and 35 percent of calories. Animal flesh, seeds, and nuts are all good sources of fat.

Proteins

4-8. Protein is the major structural component of all cells. Major functions include rebuilding muscle and repairing tissue. Proteins are made up of amino acids. Those that contain adequate proportions of all essential amino acids are considered complete proteins. Complete proteins include meat sources, dairy and eggs. While most plants do not contain complete protein, eating a variety of sources of incomplete protein will ensure the body receives the essential amino acids it needs. Foods that are not complete proteins but are still good sources of protein include beans, nuts and seeds.

---

1. Carotenoids are the phytonutrients responsible for the red and yellow colors in many plants, which also provide antioxidant benefits when consumed by humans.

Vitamins

4-9. Vitamins occur in small quantities in many foods and are essential for normal growth and health. Their chief function is to regulate the body processes. Vitamins can generally be placed into two groups: fat-soluble and water-soluble. Fat-soluble vitamins (D, E, A, K) need fat and minerals for absorption and storage in the body's fat and liver. Water-soluble vitamins (all except those listed previously) need frequent replenishing. The body only stores slight amounts of the water-soluble type. During a long isolating event where a routinely balanced diet is not available, an isolated person must overcome food aversions and eat as much of a variety of vitamin-rich foods as possible. Often one or more of the basic food groups (protein, grains, dairy, vegetables and fruits) are not available in the form of familiar foods, and vitamin deficiencies such as beriberi or scurvy result.

Minerals

4-10. There are two categories of minerals: a) major minerals and b) micro or trace minerals, both of which are essential for critical structural elements, controlling the action of nerves and muscles, maintenance of the body's water balance and the regulation of multiple essential bodily functions. Common examples of minerals include sodium, magnesium, potassium, calcium, and iron. The impact of a mineral deficiency can include poor body water regulation, compromised immune function, an increase the prevalence of infection and physical and cognitive decrements.

FOOD AVAILABILITY

4-11. Operational environments play a significant role in determining how much food an isolated person requires. In cold environments, more calories will be required to maintain necessary body heat. In hot environments, more water and less food will be required to meet their needs. Isolated personnel should learn what and how to consume and ration their food and water. If they are low on water (less than one quart a day), they should stay away from starchy, dry, salty food sources. These types of items will require more water to digest and increase thirst. Consuming foods with high carbohydrate content is best for water conservation, while high protein foods tend to require more water for digestion.

Plants

4-12. Plant foods provide carbohydrates—the main source of energy. Many plants provide enough protein to keep the body at normal efficiency. Although plants may not provide a balanced diet, they will sustain you even in the arctic, where meat's heat-producing qualities are normally essential. Many plant foods such as nuts and seeds will give you enough protein and oils for normal efficiency. Roots, green vegetables, and plant foods containing natural sugar will provide calories and carbohydrates that give the body natural energy.

4-13. The food value of plants becomes more and more important if you are eluding the enemy or if you are in an area where wildlife is scarce. For instance—

- You can dry plants by wind, air, sun, or fire. This retards spoilage so that you can store or carry the plant food with you to use when needed.
- You can obtain plants more easily and more quietly than meat. This is extremely important when the enemy is near.

Animals

4-14. Meat is more nourishing than plant food. In fact, it may even be more readily available in some places. However, to get meat, you need to know the habits of and how to capture the various wildlife. To satisfy your immediate food needs, first seek the more abundant and more easily obtained wildlife, such as insects, crustaceans, mollusks, fish, and reptiles. These can satisfy your immediate hunger while you are preparing traps and snares for larger game. General considerations for animals includes the following—

- Insects are plentiful and very easy to forage. Insects live close to the ground under logs and rocks, and in underground colonies.
- Fish are procured by the use of items such as hooks, fishing line, weights, spears, or lures.
- Knowledge of fishing techniques will give isolated persons the edge when procuring fish. Basic principles of how fish react and how to trap, prepare, and consume them will provide a plentiful source of food near fresh or saltwater areas.

4-15. Small mammals, such as rabbits, prairie dogs, and rats, often inhabit arid regions. Look in the shade or in underground burrows to protect themselves from the sun and heat. Small animals such as rodents usually live in desert caves or rock outcroppings. Smoke sometimes works to force small animals, such as rabbits, out of their burrows. Placing traps and snares at openings of these burrows is the best method to catch them.

BASIC FOOD PREPARATION

4-16. Once plants are identified as edible, they need only be washed to ensure that all potential fertilizer or other toxins are rinsed off. The foliage is then cut up into usable pieces and either consumed raw or cooked as needed.

4-17. Clean and eat sea life as soon as possible to avoid spoilage. Preserve leftover meat by drying or smoking. Use the internal parts as bait for future fishing opportunities.

4-18. Sea birds have proven to be a useful source of food. Skin freshly killed birds, rather than plucking, to remove oil glands. The meat is edible raw, sun-dried, or cooked. The gullet contents can be a good food source. Eat the flesh or preserve immediately after cleaning. The viscera, along with any other unused parts, makes good fish bait.

4-19. Game food should be fully prepared with the least amount of effort and exertion. Almost every part of game food can be consumed. Skinning and butchering should be done so that all edible meat can be saved. Cutting large animals up into smaller pieces will assist in moving the meat closer to where you are camped. When removing the skin of an animal, you must decide whether you are going to use the hide to improvise an item like a piece of clothing or a water bag, or if you will just discard the skin after it is removed: if it is to be discarded, a rough skinning job can be completed. The best time to butcher and skin an animal is right after it is killed. At that time, the animal is still hydrated and the skin will come off the easiest. The internal organs can spoil the rest of the meat if not removed in a timely fashion. If the animal is killed late in the day, the organs can be removed and the rest of the butchering can be done at a later time. An effort should be made to secure the carcass from predators.

4-20. When butchering an animal, all edible fat should be saved—especially in colder climates where an isolated person's diet might consist of only lean meat. Fat should be eaten to provide for a more balanced and complete diet.

4-21. Small game skinning of animals like squirrels, rabbits, and other small-size game begins with removing the entrails. This is done by splitting the body open and pulling them out with the fingers. This should also include the chest cavity.

4-22. During preparation, ensure that all items used for their processing and preparations like tools and cutting surfaces are cleaned well. Avoid cross contamination with other foods, improper handling of raw meat can spread bacteria and make you ill. Prevent raw juices from raw meat and seafood from dripping onto other food during the storage and preparation process. Wash hands before and after food handling to minimize the chances of spreading bacteria that could make you ill. Prepare a sanitizer by mixing one teaspoon of liquid bleach per one quart of water. Use this to wash down all contaminated surfaces. Leave the solution on the surfaces for a full ten minutes before rinsing off.

Note: There is less work and less wear on cutting implements if the ribs are broken first, then the breaks are cut.

BASIC COOKING AND PRESERVATION METHODS

4-23. Fully cook wild game, large insects, freshwater fish, clams, mussels, snails and crawfish to kill any internal parasites. Mince mussels and large snails to make them tender for consumption. Cooking these items will also improve the taste and, in some cases, allow the isolated person to stretch out how much food they have (for example, turning a small rabbit into a large stew). Cooking will also slow the decomposition of food but will not eliminate it.

4-24. In some cases, cooking food properly can produce more calories from the food, as in the case of starches. Cooked starches have more calories for use as energy than raw starches. Cooking in many cases can destroy harmful anti-nutrients that bind minerals. The destruction of these anti-nutrients can assist in the absorption of beneficial nutrient food compounds. Cooking will destroy many bacteria that are present which work to break the food down causing it to spoil faster.

4-25. Cooking methods that are the best for immediate consumption and provide the most nutrition are the least effective for food preservation. Reheat boiled and steamed food every day until fully consumed. The danger zone for food bacteria growth is between 40 and 140° F. Heat meat to an internal temperature of 180° F. When food is reheated, it should be heated to an internal temperature of at least 165° F. The symptoms of food borne illness like diarrhea or vomiting can cause severe dehydration and can be serious to the isolated person.

LEACHING

4-26. Leaching food is a technique that is used to prepare food for cooking by removing unwanted substances that are water-soluble minerals like potassium, for example. Removing bad tastes from food and softening food are other benefits of leaching. Placing cleaned and shelled acorns in a clean fast flowing stream will remove tannic acid from acorns or wild asparagus so it can then be roasted and consumed. If no stream is available, leaching can be done by soaking food in several changes of water.

BOILING

4-27. Boiling is a high heat cooking technique and is the most nutritious, simplest, and safest method of cooking (see Figure 4-1 on page 112). Boiling allows you to heat and moisten plants and meat, breaking them down and making them more easily digestible. Boiling allows all the nutrients to stay in the broth of the stew. It allows the addition of plants and water to the mixture to increase the amount of food and improve the taste. You can eat the meat and plants while hydrating and receiving nutrients from the leftover broth. Boiling works well for feeding larger groups.

4-28. Numerous containers can be used for boiling. For example, suspend a metal container above, or set beside, a heat source to boil foods. Green bamboo makes an excellent cooking container.

Parboiling

4-29. Parboiling is the partial boiling of food. Parboiling is used to remove foul-tasting substances from food. Parboiling softens foods that are going to be cooked in another way like baking or grilling.

Steaming

4-30. Steaming is an indirect heat method of cooking that works by using boiling water to produce steam and then using that steam to cook the food. For example, a metal container suspended above, or set beside, a heat source to boil foods. As with boiling, green bamboo makes an excellent cooking container.

Stone Boiling

4-31. Stone boiling is a method of boiling using superheated rocks and a container that holds water but cannot be suspended over an open flame. Examples of these containers are survival kit containers, a helmet, a hole in the ground lined with waterproof material, or a hollow log. The container is filled with food and water and then heated with super-hot stones until the water boils. Stones from a stream or damp area should not be used. The moisture in the stones may turn to steam and cause the stones to explode while they are being heated in the fire. The container should be covered and new stones added as the water stops boiling. The rocks can be removed with the aid of a wire secured to the rock before being put into the container or two sticks used in a chopstick fashion.

Figure 4-1. Boiling

BAKING

4-32. Baking is a good method of cooking as it is slow and allows the food's nutrients to stay in the container, retaining a large amount of the moisture in the food and making tough meat tender and easier to consume. Baking is often used with various types of ovens. Ovens that can be improvised in the field include the bank oven. This can be produced when you find the bank of a river or hill comprised of clay dirt. Go to the top of the bank and put a stick in the ground about two feet back from the bank drop off. This will create the hole for the chimney so the oven can breathe. Then at the bank's face, a hole will be dug for the cooking compartment with a small opening and a larger interior space. Once the hole is dug, you can build a fire inside the cooking compartment to cure it. A door can be made from a large rock or green log. Before use, heat up a set of coals, place inside the oven and begin to bake.

Reflector Oven

4-33. The reflector oven works well for baking and is easy to make. Just build a framework that will reflect the intense heat from a fire. To improve the effectiveness of the oven you can add angles to the framework and if available add a lining of aluminum foil to intensify the reflection capability. Build a fire in front of the reflector achieving a nice set of coals. Place the food on top of the coals, keep the fire stoked, and bake. Temperature control for ovens can be measured and adjusted with the use of flour made out of dried plant grains or seeds.

Baking with Leaves

4-34. Foods can be baked by wrapping in wet leaves (avoid using a type of plant that will give an unpleasant flavor to what is being cooked), placed inside a metal container, or packed with mud or clay and placed directly on the coals. Fish and birds packed in mud and baked need not be skinned because the scales, skin, or feathers will come off the animal when the mud or clay is removed.

4-35. Clambake-style baking is done by heating a number of stones in a fire and allowing the fire to burn down to coals.A layer of wet seaweed or leaves is then placed over the hot rocks. Food such as mussels and clams in their shells are then placed on the wet seaweed or leaves. More wet seaweed or leaves and soil is used as a cover. When thoroughly steamed in their juices, clam, oyster, and mussel shells will open and may be eaten without further preparation.

ROASTING

4-36. Roasting food is a less desirable method of cooking because it involves exposing food to direct heat. This can quickly destroy the nutritional properties of the food. The fire will not only dry out the food—it will also cause the food's very important juices to cook out and end up in the fire. Putting a piece of food on a stick and placing it over a fire is considered roasting. Planking is another form of roasting. This involves meat, which can be cooked on a flat surface such as a rock, a board, or a piece of bamboo. This surface is propped up next to the fire. The radiant heat from the fire cooks the meat. The meat will need to be turned over to thoroughly cook both sides.

FRYING

4-37. Frying is the least favorable method of preparing food. Nearly all of the natural juices are cooked out of the meat, which tends to make it tough, and some of the meat's nutritional value will be destroyed. Frying can be done on any nonporous surface which can be heated. Examples are unpainted metal from an aircraft, vehicle parts, large seashells, flat rocks, and some survival kits.

PRESERVING FOOD

4-38. Preserving food is very important for isolated persons. Food, especially meat, has a tendency to spoil within a short period of time, unless it is preserved. There are many ways to preserve food including cooking, refrigeration, freezing, and dehydration.

4-39. Cooling is an effective method of storing food for short periods. Heat tends to accelerate the decomposition process where cooling retards decomposition. The colder food becomes, the less likely it is to deteriorate before freezing (which eliminates decomposition). Cooling devices available include—

- **Snow**. Food items buried in snow will maintain a temperature of approximately 32° F (0° C).
- **Streams**. Food wrapped in waterproof material and placed in streams will remain cool in summer months. Take care to secure the food to prevent loss.
- **Below ground soil**. Particularly in shady areas or along streams, it remains cooler than the surface. Dig a hole, line with grass, and cover to form an effective cool storage area similar to a root cellar.
- **Water**. When water evaporates, it tends to cool down the surrounding area. Wrap Articles of food in an absorbent material such as cotton or burlap and re-wet as the water evaporates.
- **Freezing**. Once food is frozen, it will not decompose. Freeze food in meal-sized portions.
- **Drying**. Drying removes all moisture and preserves the food. Dry all food by sunning, smoking, or burying it in hot sand.

4-40. Many animals and insects will devour an isolated person's food supply if it is not correctly stored. Protect food from insects and birds by wrapping it in material, wrapping and tying brush around the bundle, then wrapping it with another layer of material. This creates "dead air" space making it more difficult for insects and birds to get to the food. If the outer layer is wetted, evaporation will also cool the food to some degree. In most cases, if food is stored several feet off the ground, it will be out of reach of most animals. Hang the food by putting it into a "cache." If the food is dehydrated, the container must be waterproof to prevent reabsorption.

4-41. Frozen food will remain frozen only if the outside temperature remains below freezing. Burying food is a good method of storage as long as scavengers are not in the area to uncover it. Consider insects and small animals when burying food. Food should never be stored in your shelter as this may attract wild animals and could be hazardous to you.

SUN DRYING

4-42. To sun dry food, cut it into thin slices so it is less desirable to insects and place it in direct sunlight. Cut meat across the grain and remove all of the fat to improve tenderness and decrease drying time. Add salt to improve taste and accelerate the drying process. Decrease drying time by stringing the meat on a rack so that it dries using the sun and the wind.

SMOKING

4-43. Use a non-resinous hard wood such as willow or aspen to produce smoke that will provide slow heat to dry the meat and add flavor. Cut the meat into small, thin slices with all fat removed. Place the meat over a set of heated coals on an improvised rack about two feet above the coals. The goal is to produce heat and smoke—not flame. Place some type of material, such as a poncho or a piece of nylon, around the meat to contain the smoke. Place chips of green wood on the coals to produce the smoke that will preserve the meat. Meat smoked overnight in this manner will last up to a week. Two days of continuous smoking will preserve the meat for up to four weeks. You can consume properly-smoked meat; it will look like a dark, curled, brittle stick, without further cooking.

4-44. If the situation allows, the isolated person needs to gather and preserve more plant foods than can be eaten, preserving excess in the same manner as animal foods. Dry plants and fruits by wind, air, sun, or fire, with or without smoke. Any combination of these methods will remove moisture. If the plant part is large, such as tubers, it should be sliced and then dried. Some type of protection may be necessary to prevent consumption or contamination by insects. Carry extra fruits or berries by wrapping them in leaves or moss.

PLANTS

4-45. If personnel enter into an isolating event with some basic knowledge and information, they will be able to supplement or totally survive off plant life in the area if it is abundant and available. Plants are valuable sources of food because they are widely available, easily procured and, in the proper combinations, can meet all nutritional needs. Plants provide carbohydrates, which provide the body energy and calories. Carbohydrates keep weight and energy up and include important starches and sugars. Many types of vitamins and minerals can be found in different varieties of plants and can provide the crucial nutrients that the body needs to remain healthy and disease free.

4-46. If the plant includes an inedible or poisonous variety in its family, the edible plant must be distinguishable from the poisonous one to the average eye. Personnel can find valuable information on edible plants in the area of operations by doing a country study or purchasing a plant identification book or set of plant identification cards. There is also plant information on the evasion chart personnel can procure.

4-47. The best and safest method for edibility of a plant is positive identification of the plant. If positive identification is not possible, then use the edibility test. When selecting plant foods, the following should be considered:

- Look for plant parts that contain high energy such as fruit, seeds, grains, nuts, tubers, and bulbs.
- Select plants resembling those cultivated by people.

4-48. It is risky to rely on plants being edible for human consumption simply because animals have also been eating the plants. The animal may be able to digest certain toxins that a person cannot. When selecting an unknown plant as a possible food source, apply the following rules:

- Often plants growing near homes and occupied buildings or along roadsides have been sprayed with pesticides. Wash these plants thoroughly. In more highly developed countries with many automobiles, avoid roadside plants, if possible, due to contamination from exhaust emissions.
- Boil or disinfect plants growing in contaminated water.
- Some plants develop extremely dangerous fungal toxins. To lessen the chance of accidental poisoning, do not eat any fruit starting to spoil or showing signs of mildew or fungus.
- Plants of the same species may differ in their toxic or sub-toxic compounds content because of genetic or environmental factors. One example of this is the foliage of the common chokecherry. Some chokecherry plants have high concentrations of deadly cyanide compounds but others have low concentrations or none. Horses have died from eating wilted wild cherry leaves. Avoid any weed, leaves, or seeds with an almond-like scent—a characteristic of the cyanide compounds.

- Some people are more susceptible to gastric distress (from plants) than others. If personnel are sensitive in this way, they should avoid unknown wild plants. If they are extremely sensitive to poison ivy, they should avoid products from this family, including any parts from sumacs, mangoes, and cashews.

- Some edible wild plants, such as acorns and water lily rhizomes, are bitter. These bitter substances, usually tannin compounds, make them unpalatable. Boiling them in several changes of water will usually remove these bitter properties.

- Many valuable wild plants have high concentrations of oxalate compounds, also known as oxalic acid. Oxalates produce a sharp burning sensation in the mouth and throat and damage the kidneys. Baking, roasting, or drying usually destroys these oxalate crystals. The corm (bulb) of the jack-in- the-pulpit is also known as the "Indian turnip," but it can be eaten after removing these crystals by slow baking or by drying.

WARNING: Avoid mushrooms and fungi; they are not technically plants. Fungi have toxic peptides, a protein-based poison that has no taste. They are of little nutritional value and it is very difficult to determine if they are poisonous or safe to consume.

There is no field test for determining if an unknown mushroom or fungus is edible or deadly except by eating it. There are many poisonous varieties, and some wild mushrooms are hard even for experts to identify. The lack of nutrients and the massively increased risk of accidental poisoning make them a nonviable food source.

- Avoid plants with umbrella-shaped flowers although carrots, celery, dill, and parsley are members of this family (see Figure 4-2 on page 118). One of the most poisonous plants, water hemlock, is also a member of this family.

Figure 4-2. Umbrella-Shaped Flowers

4-49. Avoid all of the following—

- All of the legume family (beans and peas). They absorb minerals from the soil and can cause digestive problems. Selenium is the most common mineral absorbed is which has given locoweed its fame. (Locoweed is a vetch, which is an herbaceous plant of the pea family).

- Plants with a milky sap; this is an indicator of a poisonous plant. Do not eat plants that are irritants to the skin, such as poison ivy.

4-50. Examples of poisonous bulbs include tulips and death camas. Also—

- Avoid white and yellow berries for they are frequently poisonous. About one-half of all red berries are poisonous. Blue or black berries are generally safe for consumption.

- Consider plants with shiny leaves as poisonous.

- The following are generally considered safe to eat:
  - Aggregated fruits and berries (for example, thimbleberry, raspberry, salmonberry, and blackberry) are always edible.

- Single fruits on a stem.

PLANT IDENTIFICATION

4-51. In addition to memorizing particular varieties through familiarity, plants may be identified by using such factors as—

- Leaf shape and margin.
- Leaf arrangements.
- Root structure.

4-52. The basic leaf margins are toothed, lobed, and toothless or smooth (see Figure 4-3).

Figure 4-3. Leaf Margins

Toothed   **Toothless**   **Lobed**

4-53. Leaves may be lance-shaped, elliptical, egg-shaped, oblong, wedge-shaped, triangular, long-pointed, or top-shaped (see Figure 4-4 on page 120).

Figure 4-4. Leaf Shapes

Lance-shaped

Elliptical

Egg-shaped

Oblong

Wedge-shaped

Triangular

Long-pointed

Top-shaped

4-54. As shown in Figure 4-5 on page 121, the basic types of leaf arrangements are opposite, alternate, compound, simple, and basal rosette.

Figure 4-5. Leaf Arrangements

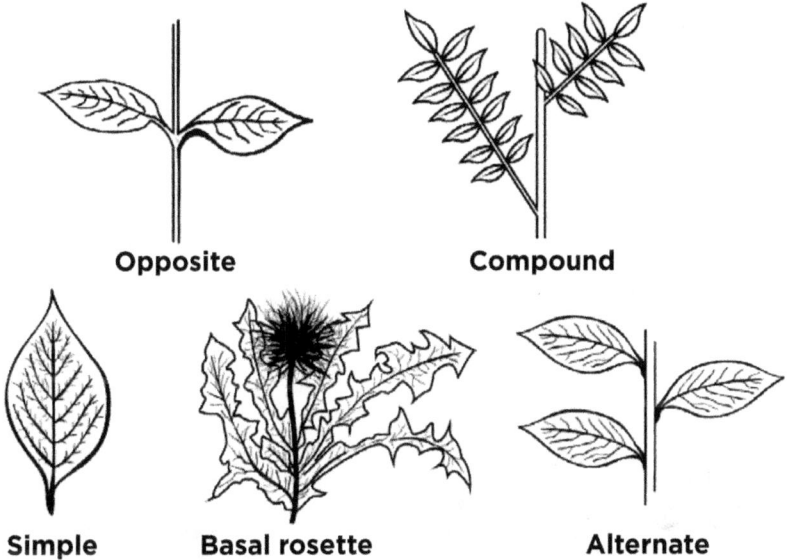

Opposite　　　　　**Compound**

Simple　　**Basal rosette**　　　　**Alternate**

4-55. The basic types of root structures, shown in Figure 4-6 on page 122, are the taproot, tuber, bulb, rhizome, clove, corm, and crown. Bulbs (onions), when sliced in half, will show concentric rings. Cloves are those bulb-like structures that remind us of garlic and will separate into small pieces when broken apart. This characteristic separates wild onions from wild garlic. Taproots resemble carrots and may be single-rooted or branched, but usually only one plant stalk arises from each root. Examples of tubers are potatoes and daylilies. You will find these structures either on strings or in clusters underneath the parent plants. Rhizomes are large, creeping rootstock or underground stems. Many plants arise from the "eyes" of these roots. Corms are similar to bulbs but are solid when cut rather than possessing rings. A crown is the type of root structure found on plants such as asparagus. Crowns look much like a mop head under the soil's surface.

Figure 4-6. Root Structures

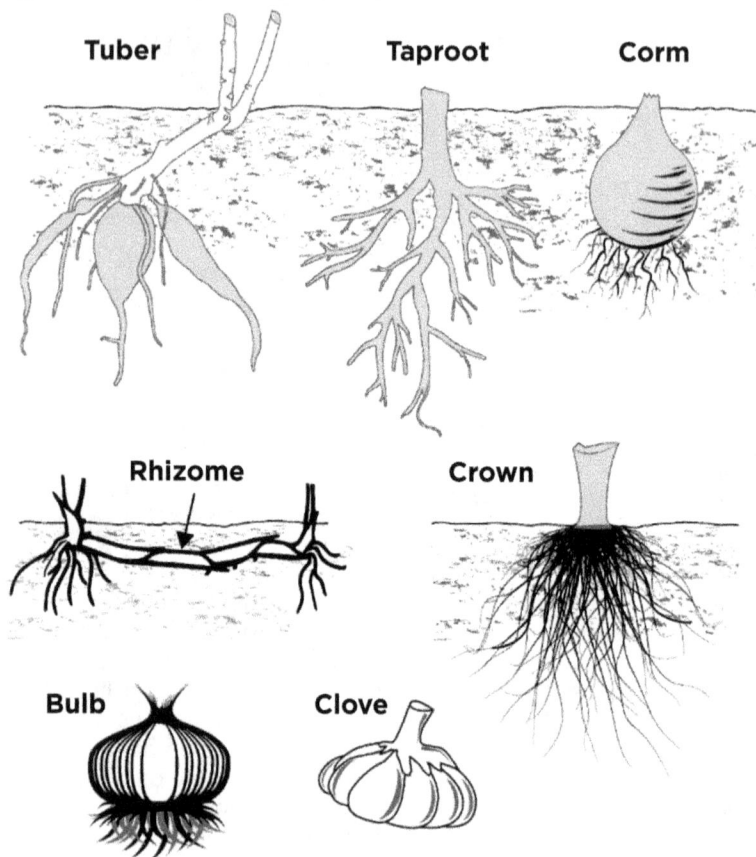

Tuber   **Taproot**   **Corm**

Rhizome   **Crown**

Bulb   **Clove**

4-56. Learn as much as possible about the unique characteristics of plants used for food. Some plants have both edible and poisonous parts. Many are edible only at certain times of the year. Others may have poisonous relatives that look very similar to the edible varieties of those used for medicinal purposes.

PLANT EDIBILITY TESTING

4-57. Select plants that grow in sufficient quantity within the local area to justify the edibility test and provide a lasting source of food if the plant proves edible. Plants growing in water or moist soil are often the most palatable. Plants growing in shaded areas are less bitter. There are exceptions to every rule, but isolated persons should only select unknown plants as a last resort.

4-58. When selecting unknown plants for possible consumption, remember the poisonous characteristics to avoid. Apply the edibility test to only one plant at a time so if some abnormality does occur, it will be obvious which plant caused the problem. Once a plant has been selected to be tested, proceed as follows:

Step 1 If there are any unpleasant odors such as a moldy or musty smell coming from the plant, stop testing and disregard as a possible edible plant option. Also, if the plant gives off an "almond" scent, disregard it as a possible edible plant option.

Step 2 Crush or break part of the plant to determine the color of its sap. If the sap is clear, proceed to the next step.

Step 3 Touch the plant's sap or juice to the inner forearm. If there are no ill effects, such as a rash or burning sensation to the skin, then proceed with the rest of the steps.

Step 4 If a there was not an ill reaction when touching the inner forearm, place some of the plant juice on the outer lip for eight minutes. If a reaction occurs, stop the test.

Step 5 If still no reaction, taste a small pinch of the plant and leave it in the mouth for eight minutes. If there is an unpleasant taste, such as bitterness or a numbing sensation of the tongue or lips, stop the test and spit out the plant. If a reaction does not occur, swallow the pinch of plant.

Note: Sometimes heavy smokers are unable to taste various poisons, such as alkaloids.

Step 6 After swallowing, wait eight hours. If there is no reaction after eight hours, chew a handful of the plant, swallow, and wait an additional eight hours. If no reaction occurs after eight hours, consider the tested plant part edible.

Step 7 Eat any new or strange food with restraint until the body has become accustomed to it. The plant may be slightly toxic and harmful when eaten in large quantities.

CAUTION: Test all parts of the plant for edibility, as some plants have both edible and inedible parts. Do not assume that a part that proved edible when cooked is also edible when raw. Test the part raw to ensure edibility before eating raw. The same part or plant may produce varying reactions in different individuals.

4-59. Before testing a plant for edibility, make sure there are enough plants to make the testing worth the time and effort. Each part of a plant (roots, leaves, flowers, and so on) requires more than 24 hours to test. Do not waste time testing a plant that is not relatively abundant in the area.

4-60. Remember that eating large portions of plant food on an empty stomach may cause diarrhea, nausea, or cramps. Two good examples of this are such familiar foods as green apples and wild onions. Even after testing plant food and finding it safe, eat it in moderation.

4-61. Testing for edibility involves considerable time, so it is important to be proficient at identifying edible plants. To avoid potentially poisonous plants, stay away from any wild or unknown plants that have—

- Milky or discolored sap.
- Beans, bulbs, or seeds inside pods.
- A bitter or soapy taste.
- Spines, fine hairs, or thorns.
- Foliage that resembles dill, carrot, parsnip, or parsley.
- An almond scent in woody parts and leaves.
- Grain heads with pink, purplish, or black spurs.
- A three-leafed growth pattern.

4-62. Using the above criteria as eliminators when choosing plants for the universal edibility test will cause isolated persons to avoid some edible plants. More important, these criteria will often help in avoiding plants that are potentially toxic to eat or touch. Personnel should learn as much as possible about the plant life of the areas where they train regularly and where they expect to be operating. The following are ten globally available plants to identify and use. These plants are in almost every biome:

- Cattail.
- Segmented grass.
- Dandelion.
- Aggregate berries.
- Thistle.
- Onions.
- Clover.
- Plantain.
- Wild rose hip.
- Fiddlehead fern.

4-63. The following are plants that are safe sources for food:

- Amaranth (Amaranths retroflex and other species).
- Arrowroot (Sagittarius species).
- Asparagus (Asparagus officinalis).
- Beechnut (Fagus species).
- Blackberries (Rubes species).
- Blueberries (Vaccinium species).
- Burdock (Arctium lappa).
- Chestnut (Castanea species).
- Chicory (Cichorium intybus).

- Chufa (Cyperus esculentus).
- Daylily (Hemerocallis fulva).
- Nettle (Urtica species).
- Oaks (Quercus species).
- Persimmon (Diospyros virginiana).
- Pokeweed (Phytolacca Americana).
- Prickly pear cactus (Opuntia species).
- Purslane (Portulaca oleracea).
- Sassafras (Sassafras albidum).
- Sheep sorrel (Rumex acetosella).
- Strawberries (Fragaria species).
- Water lily and lotus (Nuphar, Nelumbo, and other species).
- Wild onion and garlic (Allium species).
- Wild rose (Rosa species).
- Wood sorrel (Oxalis species).
- Bamboo (Bambusa and other species).
- Bananas (Musa species).
- Breadfruit (Artocarpus incisa).
- Cashew (Anacardium occidental).
- Coconut (Cocoa nucifera).
- Mango (Mangifera indica).
- Palms (various species).
- Papaya (Carica species).
- Sugarcane (Saccharum officinarum).
- Taro (Colocasia species).
- Acacia (Acacia farnesiana).
- Agave (Agave species).
- Cactus (various species).
- Date palm (Phoenix dactylifera).
- Desert amaranth (Amaranths palmeri).

GLOBALLY-COMMON EDIBLE PLANTS

4-64. Cattail. Underground stems (rhizomes) are high in starch and can be gathered year round. They can be eaten raw by chewing the central starchy core after the spongy surrounding tissue is removed. When boiled the entire rhizome can be eaten. Young shoots, green flowers, yellow pollen and the white inner leaves of immature stems in spring and early summer are also provide nutrition to the isolated person.

4-65. Palms. Globally there are approximately 2,600 species found in temperate, desert, and tropical climates. Palm hearts (the part of the plant or tree form which the leaves branch) are often sweet, containing starch that provide energy. Also, palm fruit including coconuts and dates are edible.

4-66. Ferns. There are over 10,000 species globally with only a couple considered mildly toxic. Fiddleheads, the young unrolling fern fronds, have the appearance of the end of a violin. You can eat Fiddleheads raw although they taste better cooked. Rhizomes store starch energy reserves and many ferns can be used to expel stomach and intestinal worms.

4-67. Bamboo. Over 1200 species are found globally and is the largest of all grasses. Young shoots (rhizomes) are edible and have the most nutritional value, use leaves and stems as fillers. All other grasses are edible.

SEAWEED

4-68. Seaweed is a form of marine algae found on or near ocean shores. There are also some edible freshwater varieties. Seaweed is a valuable source of iodine, other minerals, and vitamin C. Large quantities of seaweed in an unaccustomed stomach can produce a severe laxative effect. The following lists various types of edible seaweed:

- Dulse (Rhodymenia palmata).
- Green seaweed (Ulva lactuca).
- Irish moss (Chondrus crispus).
- Kelp (Alaria esculenta).
- Laver (Porphyra species).
- Mojaban (Sargassum fulvellum).
- Sugar wrack (Laminaria saccharina).

4-69. When gathering seaweed for food, find living plants attached to rocks or floating free. Seaweed washed onshore any length of time may be spoiled or decayed. Dry freshly-harvested seaweed for later use.

4-70. Different types of seaweed are prepared in different ways. Thin and tender varieties of seaweed are sun dried or over a fire until crisp, then crushed and added to soups or broths. Boil thick, leathery seaweeds for a short time to soften them and eat as a vegetable or with other foods. Some varieties can be consumed raw after testing for edibility.

PREPARING PLANTS FOR CONSUMPTION

4-71. Although some plants or plant parts are edible raw, others must be cooked to be edible or palatable. Edible means that a plant or food will provide necessary nutrients to an isolated person. Palatable means that it is pleasing to eat. Many wild plants are edible but barely palatable. It is a good idea to learn to identify, prepare, and eat wild foods.

4-72. Methods used to improve the taste of plant food include soaking, boiling, cooking, or leaching. Leach by crushing the food (for example, acorns), placing it in a strainer, and pouring boiling water through it or immersing it in running water.

4-73. Boil leaves, stems, and buds until tender, changing the water, if necessary, to remove any bitterness.

4-74. Boil, bake, or roast tubers and roots. Drying helps to remove caustic oxalates from some roots like those in the arum family.

4-75. Leach acorns in water, if necessary, to remove the bitterness. Some nuts, such as chestnuts, are good raw—but taste better roasted.

4-76. You can consume many grains and seeds raw until they mature. When they are hard or dry, they may need to be boiled or ground into meal or flour.

4-77. The sap from many trees, such as maples, birches, walnuts, and sycamores, contains sugar. Boil these saps to make syrup for sweetening. It takes about 9 gallons of maple sap to make 1 quart of maple syrup.

MAMMALS

4-78. There are at least 5,360 different species of mammals in the world. Mammals are classified under the following nine main categories (with examples):

- Egg laying mammals (spiny anteaters).
- Marsupials (kangaroo).
- Insect-eating (rats).
- Chiropetra (bats).
- Hooved animals (zebras).
- Aquatic mammals (dolphins).
- Carnivora (lions).
- Proboscidea (elephants).
- Primates (humans).

4-79. All mammals are edible. However, the polar bear and the bearded seal have toxic amounts of vitamin A in their livers. Mammals are excellent sources of protein and essential amino acids sufficient for the maintenance of life and the repair of damaged tissue. They are the best tasting food source and closest to the normal American diet. There are some drawbacks to obtaining wild mammals as a food source. For example, in a denied area, the enemy may detect any traps and snares used to catch the animals.

4-80. Take caution, mammals have teeth and will bite in self-defense. The bite from a mammal can cause a severe infection. The amount of injury that a mammal can inflict is in direct proportion to its size. Any mother mammal can be extremely aggressive in defense of her young. Any mammal with no route of escape will fight when cornered. The duck-billed platypus, native to Australia and Tasmania, is an egg-laying, semi-aquatic mammal that has poisonous claws on its hind legs. Scavenging mammals, such as the coyote, may carry diseases. Isolated persons must be prepared to hunt, shoot, trap, or snare mammals from their environment.

HUNTING

4-81. To become successful in hunting, isolated persons must go through a behavioral change and reorganize personal priorities. This means the one and only goal for the present is to kill an animal for food. To kill this animal, you must mentally become a predator. You must be willing and prepared to undergo stress to hunt down and kill an animal. Because of the type of field-expedient weapons isolated persons are likely to have, it will be necessary to get very close to the animal to immobilize or kill it. This requires considerable stealth and cunning. Knowledge of the animal is also very important. Observe the animal life of the area by studying signs such as trails, droppings, and bedding areas. Isolated persons should understand the following general characteristics of the animals in the area:

- The size of the tracks gives a good idea of the size of the animal.
- The depth of the tracks indicates the weight of the animal.
- Animal droppings provide much information. For example, if it is still warm or slimy, it was made very recently; if there is a large amount scattered around the area, it could well be a feeding or bedding area. The droppings may also indicate what the animal feeds on. Carnivores often have hair and bone in their droppings. Herbivores have coarse portions of the plants they have eaten in their droppings.
- Many animals mark their territory by urinating or scraping areas on the ground or trees. These signs could indicate good trap or ambush sites.

4-82. By following those signs, (tracks and droppings), study the feeding, watering, and resting areas of the hunted animals. Well-worn trails often lead to the animal's watering place. Having made a careful study of all the animal signs, you are in a much better position to capture it, whether electing to stalk, trap, snare, or lie in wait to shoot it. Wild animals rely entirely upon their senses for their preservation.

These senses are smell, vision, and hearing. Humans have lost the keenness of some of their senses such as smelling and hearing.

4-83. To overcome this loss, humans have the ability to reason. As an example, some animals have a fantastic sense of smell, but by approaching, the quarry from a downwind direction will allow you to circumvent their keen sense of smell. The best times to hunt are at dawn and dusk as animals are either leaving or returning to their bedding areas. Both diurnal (primarily active during daylight hours) and nocturnal (primarily active at night-time) animals are active at this time. The following are five basic methods of hunting:

- **Ambush**. Ambush is the best method for inexperienced hunters as it involves less skill. The main principle of this method is to wait along a well-used game trail until the quarry approaches within killing range. Morning and evening are usually the best times to hunt. Take care not to disturb the area; always wait downwind. Patience and self-control are necessary to remain motionless.

- **Stalking**. Stalking refers to the stealthy approach toward game. When an animal has been sighted you should proceed to close the distance using all available cover. Move slowly in order to minimize noise. Quick movements are easily detected by animals. Always approach from the downwind side and move when the animal's head is down eating, drinking, or looking in another direction. Blind stalking uses the same techniques as in regular stalking, the main difference being that you are stalking a position where the animal is expected to be while the animal is not in sight.

- **Tracking**. Tracking is very difficult unless conditions are ideal. This method involves reading all of the signs left behind by the animal, interpreting what the animal is doing, and determining how it can best be killed. The most common signs are trails, beds, urine, droppings, blood, tracks, and feeding signs.

- **Funneling**. Some wild animals can be scared or driven in a direction where other persons or traps have been positioned. Using this method to funnel the animals into a valley or canyon is a good place to make a drive. More than one person is usually necessary to make a drive.

4-84. Once you have found an animal, you will need a method of killing the animal. If a firearm is available and firing it will not reveal your position to the enemy, shoot the animal from a safe distance with the firearm. Isolated persons cannot afford to waste ammunition on moving game, or game that is beyond the effective range of the firearm. Wait for the animal to pause and then place an accurate shot to a vital area on the animal. You should aim for the brain, spine, lungs, or heart. A hit in these areas is usually fatal. Take the best shot at the broad side of the animal; this also gives a larger point of aim. The shot will spoil any meat around the entry and exit wound; aim for a spot that will spoil the least amount of meat.

4-85. Some larger species of mammals have very thick skulls, and bullets may not fully penetrate through the skull plate. Larger animals may not go down right away; therefore, once the shot is taken, you should be prepared to track the animal. You may have to follow the animal's track or blood trail until you find the animal. Ensure the animal is dead before approaching the animal, as it may attack when wounded. One can poke the animal in the eye with a large stick to ensure it is safe to approach. Hunting at night will allow the hunter to get closer to animals. A bright flashlight or an improvised torch can blind the animal. If a firearm is not available, an improvised impact weapon, such as a club or a spear, may be used.

TRAPPING AND SNARES

4-86. When an isolated person is unarmed or the sound of a rifle shot could be a problem, trapping and snaring is a good alternative. Several well-placed traps have the potential to catch much more game than one person with a rifle is likely to shoot. Traps and snares also allow isolated persons to hunt while they are completing other tasks necessary for recovery.

4-87. Traps and snares work twenty-four hours a day without anyone needing to be present. Noise from animals attempting to free themselves from a trap is a consideration on how frequently the trap is checked, as well as its placement. A periodic check of the traps and game retrieval, and/or trap resets are required for a successful hunt. For any type of trap or snare to be effective, you must—

- Be familiar with the species of animal you intend to catch.
- Be capable of constructing a proper trap and properly masking your scent.
- Not alarm the prey by leaving signs of human presence.

4-88. Isolated persons must determine what species of game are in the area and set traps specifically for those animals in mind. Look for the following:

- Runs and trails.
- Tracks.
- Droppings.
- Chewed or rubbed vegetation.
- Nesting or rooting sites.
- Feeding and watering areas.

4-89. Position traps and snares where there is proof that animals pass through; distinguish between a "run" and a "trail." A trail will show signs of use by several species and will be rather distinct. A run is usually smaller and less distinct and will only contain signs of one species. You might construct a perfect snare, but it will not catch anything if it is haphazardly placed in the woods. Animals have bedding areas, water holes, and feeding areas with trails leading from one to another; place snares and traps around these areas to be effective. In a denied area, trap and snare concealment is important. However, it is equally important not to create a disturbance that will alarm the animal and cause it to avoid the trap.

4-90. If you must dig, move all fresh dirt away from the area. Most animals will instinctively avoid a pitfall-type trap. Prepare the various parts of a trap or snare away from the site, carry them in, and set them up. Such actions make it easier to avoid disturbing the local vegetation, thereby alerting the prey. Do not use freshly cut, live vegetation to construct a trap or snare. Freshly cut vegetation "bleeds sap" that emits an odor smelled by the prey. It is an alarm signal to the animal.

4-91. Remove or mask the human scent on and around set traps. Although birds do not have a developed sense of smell, nearly all mammals depend on smell even more than on sight. Even the slightest human scent on a trap alarms the prey and causes it to avoid the area. Actually removing the scent from a trap is difficult, but masking it is relatively easy. Use the fluid from the gall and urine bladders of previous kills. Do not use human urine. Mud, particularly from an area with plenty of rotting vegetation, is also effective. Use it to coat hands when handling the trap and to coat the trap when setting it.

4-92. In nearly all parts of the world, animals know the smell of burned vegetation and smoke. It is only when a fire is actually burning that they become alarmed. Therefore, smoking the trap parts is an effective means to mask human scent. If one of the above techniques is not practical, and if time permits, allow a trap to weather for a few days and then set it. Do not handle a trap while it is weathering. When positioning the trap, camouflage it as naturally as possible to prevent detection by the enemy and to avoid alarming the prey.

4-93. Traps or snares placed on a trail or run should use funneling or channelization. To build a channel, construct a funnel-shaped barrier extending from the sides of the trail toward the trap, with the narrowest part nearest the trap. Attempt to stay off the trail or run and construct funneling from the sides in order to limit the amount of scent on the trap or snare site. Channelization should be inconspicuous to avoid alerting the prey. As the animal gets to the trap, it cannot turn left or right and continues into the trap. Few wild animals will back up, preferring to face the direction of travel. Channelization does not have to be an impassable barrier; it only has to be inconvenient for the animal to go over or through the barrier. For best effect, the channelization should reduce the trail's width to just slightly wider than the targeted animal's body. Maintain this constriction at least as far back from the trap as the animal's body length, and then begin widening toward the mouth of the funnel.

Use of Bait

4-94. Baiting a trap or snare increases the chances of catching an animal. When catching fish, it is necessary to bait nearly all the devices. Success with a non-baited trap depends on its placement in a good location. A baited trap can draw animals in. The bait should be something the animal knows. However, this bait should not be so readily available in the immediate area that the animal can get it close by. For example, baiting a trap with corn in the middle of a cornfield would not likely work. Likewise, if corn is not grown in the region, a corn-baited trap may arouse an animal's curiosity and keep it alerted while it ponders the strange food. Under such circumstances, it may not go for the bait.

4-95. An example of bait that works well on small mammals is the peanut butter from a meal, ready-to-eat ration. Salt is also good bait. When using such bait, scatter bits of it around the trap to give the prey a chance to sample it and develop a craving for it. The animal will then overcome some of its caution before it gets to the trap. If the trap is set and baited for one species but another species takes the bait without being caught, try to determine what the animal was. Then set a proper trap for that animal, using the same bait. Scavenger-type animals may be attracted to remains of previous kills or leftover entrails; these are excellent bait for fishing.

Construction of Traps and Snares

4-96. Traps are designed to catch and hold or to catch and kill. Snares are traps that incorporate a noose to accomplish either function.

Triggers for Traps and Snares

4-97. Traps and snares crush, choke, hang, or entangle the prey. A single trap or snare will commonly incorporate two or more of these principles. The mechanisms that provide power to the trap are usually very simple. The struggling victim, the force of gravity or a bent sapling's tension provides the power.

4-98. The heart of any trap or snare is the trigger. When planning a trap or snare, ask yourself how it should affect the prey, what is the source of power, and what will be the most efficient trigger. Your answers will help you devise a specific trap for a specific species. Triggers for trapping mechanisms should be made out of dead seasoned wood that is solid in composition. Avoid green or rotted wood, which may fail, bend or break at a crucial point once the mechanism is set.

4-99. Triggers can be classified as contact triggers that are tripped by the game walks into the trap or snare, brushing the trigger and tripping the mechanism. Baited triggers put an attractant like meat, peanut butter, or other sweet or aromatic food material that will draw the game into the trap to activate the trigger system. A simple trigger can be made from the natural "Y"s of tree branches that create a natural hook that will hold another Y branch in place with friction that can contain the power of the mechanism until the game moves, tripping the trigger and releasing the mechanism activating the trap or snare. Simple triggers can also be made by carving wood to create notches in two pieces of wood that will create two hooks that interconnect the same as the "Y"s in the tree branches.

4-100. Triggers can be created utilizing a toggle that is set and held in place, creating a leverage point such that when the game passes through (in either direction) the trigger will be tripped. An example of this type of trigger is the H bi-directional trigger system. The H bi-directional trigger is made by driving two stakes in the ground with enough space for the game to pass through. Cut a notch at the top end of the two stakes— they should be cut on opposing sides of the stakes. Place a toggle that has notches that match up to the stakes, forming an "H" shape. Secure a power source like a spring pole and a slip loop to the toggle. You can also make a trip line to activate a box style trap with improvised cordage. A prop stick holding up a box style trap tied to a trip line is simple and effective: when the game hits the trip line in any direction the box will fall, trapping the game.

4-101. Leave the bark on the trigger materials used. Camouflage all exposed areas without bark that show white wood with mud. Construct carefully, check that the parts of the trigger fit, and release effectively. The components of the trigger system can be smoked over a smudge fire to de-scent them before being put in place (though do not burn the wood: the scent of burnt wood will generally alarm the game).

4-102. The following are common types of snares that might be devised:

- **Simple loop snare.** A simple snare consists of a noose placed over a trail or den hole and attached to a firmly planted stake. If the noose is some type of cordage placed upright on a game trail, use small twigs or blades of grass to hold it up. Filaments from spider webs are excellent for holding nooses open. Make sure the noose is large enough to pass freely over the animal's head. As the animal continues to move, the noose tightens around its neck. The more the animal struggles, the tighter the noose gets. This type of snare usually does not kill the animal. Cordage may loosen enough to slip off the animal's neck; therefore, wire is the best choice for a simple snare (see Figure 4-7 on page 134).

Figure 4-7. Simple Loop Snare

Twist entire length of noose for added strength and to allow locking loop to grap onto ridges.

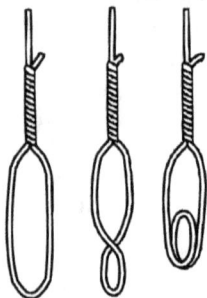

Locking mechanism

- **Apache foot snare**. The apache foot snare is an example of a hold-type trap. It is used for large browsers and grazers like deer (see Figure 4-8). This snare should be located along game trails, where an obstruction, such as a log, blocks the trail. When animals jump over this obstruction, a very shallow depression is formed where their hooves land. The Apache foot snare should be placed at this depression.

Figure 4-8. Apache Foot Snare

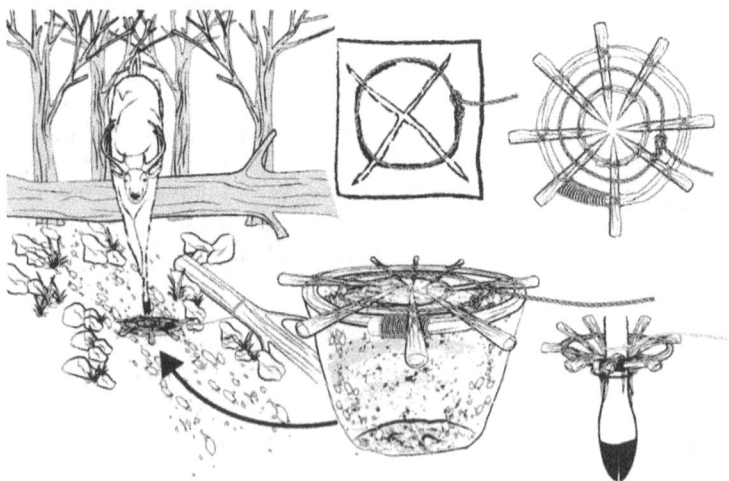

- **Drag noose snare**. Use a drag noose on an animal run. Place forked sticks on either side of the run and lay a sturdy cross-member across them. Tie the noose to the cross member and hang it at a height above the animal's head (nooses designed to catch by the head should never be low enough for the prey to step into with a foot). As the noose tightens around the animal's neck, the animal pulls the cross-member from the forked sticks and drags it along. The surrounding vegetation quickly catches the cross-member, and the animal becomes entangled (see Figure 4-9).

Figure 4-9. Drag Noose Snare

Double wire locking loop   **Channelization**

- **Twitch-up snare**. A twitch-up is a supple sapling that, when bent over and secured with a triggering device, will provide power to a variety of snares. Select a hickory or other hardwood sapling along the trail. A twitch-up will work much faster and with more force if all branches and foliage are removed. A simple twitch-up snare uses two forked sticks, each with a long and a short leg. Bend the twitch-up and mark the trail below it. Drive the long leg of one forked stick firmly into the ground at that point. Ensure that the cut on the short leg of this stick is parallel to the ground. Tie the long leg of the remaining forked stick to a piece of cordage secured to the twitch- up. Cut the short leg so that it catches on the short leg of the other forked stick. Extend a noose over the trail. Set the trap by bending the twitch-up and engaging the short legs of the forked sticks. When an animal catches its head in the noose, it pulls the forked sticks apart, allowing the twitch- up to spring up and hang the prey (see Figure 4-10).

Note: Do not use green sticks for the trigger. The sap that oozes out could glue the moving parts together.

Figure 4-10. Twitch-up Snare

- **Squirrel pole noose**. A squirrel pole is a long pole placed against a tree in an area showing a lot of squirrel activity (see Figure 4-11 on page 137). Place several wire nooses along the top and sides of the pole so that a squirrel trying to go up or down the pole will have to pass through one or more of them. Position the nooses (2~2¼ inches in diameter) about 1 inch off the pole. Place the top and bottom wire nooses 18 inches from the top and bottom of the pole to prevent the squirrel from getting its feet on a solid surface and chewing through the wire. Squirrels are naturally curious. After an initial period of caution, they will try to go up or down the pole and will be caught in the noose. The struggling animal will soon fall from the pole and strangle. Other squirrels will be drawn to the commotion. In this way, several squirrels can be caught. Multiple poles can be placed to increase the catch.

Figure 4-11. Squirrel Pole Noose

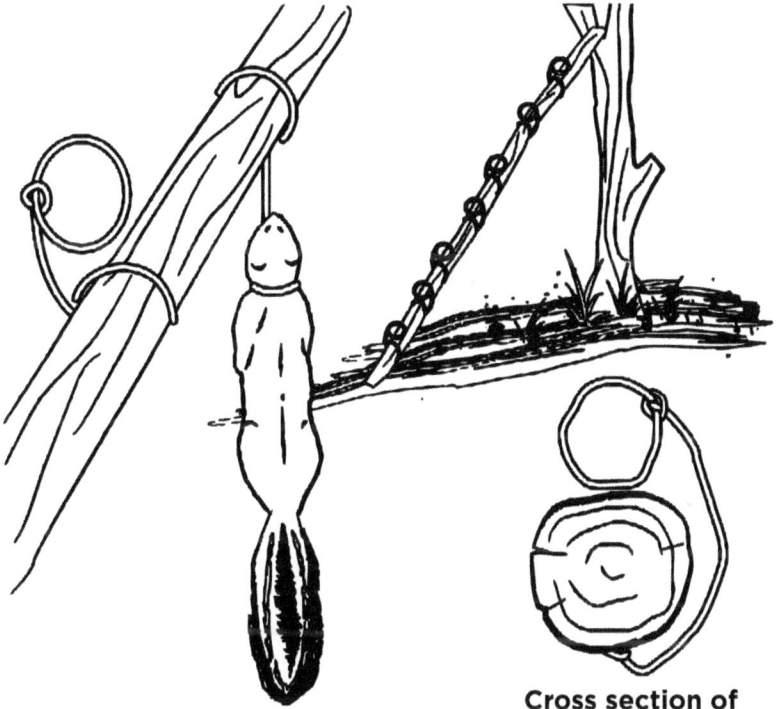

Cross section of pole and snare wire

- **Noosing wand**. A noose stick or "noosing wand" is useful for capturing roosting birds or small mammals. It requires a patient operator. This wand is more a weapon than a trap. It consists of a pole (as long as you can effectively handle) with a slip noose of wire or stiff cordage at the small end. To catch an animal, the noose is slipped over the neck of a roosting bird and pulled tight. It can also be placed over a den hole as you hide in a nearby blind. When the animal emerges from the den, jerk the pole to tighten the noose and thus capture the animal. A stout club may be needed to kill the prey (see Figure 4-12 on page 138).

Figure 4-12. Noosing Wand

- **Treadle spring snare**. Use a treadle snare against small game on a trail. Dig a shallow hole in the trail. Then drive a forked stick (fork down) into the ground on each side of the hole on the same side of the trail. Select two straight sticks that span the two forks. Position these two sticks so that their ends engage the forks. Place several sticks over the hole in the trail by positioning one end over the lower horizontal stick and the other on the ground on the other side of the hole. Cover the hole with enough sticks so that the prey must step on at least one of them to set off the snare. Tie one end of a piece of cordage to a twitch-up or to a weight suspended over a tree limb. Bend the twitch-up or raise the suspended weight to determine where to tie the trigger. The trigger should be about 2 inches long. Form a noose with the other end of the cordage. Route and spread the noose over the top of the sticks over the hole. Place the trigger stick against the horizontal sticks and route the cordage behind the sticks so that the tension of the power source will hold it in place. Adjust the bottom horizontal stick so that it will barely hold against the trigger. As the animal places its foot on a stick across the hole, the bottom horizontal stick moves down, releasing the trigger and allowing the noose to catch the animal by the foot. Because of the disturbance on the trail, an animal will be wary; therefore, channelization should be used. To increase the effectiveness of this trap, a small bait well may be dug into the bottom of the hole with bait placed in it to lure animals to the snare (see Figure 4-13 on page 139).

Figure 4-13. Treadle Spring Snare

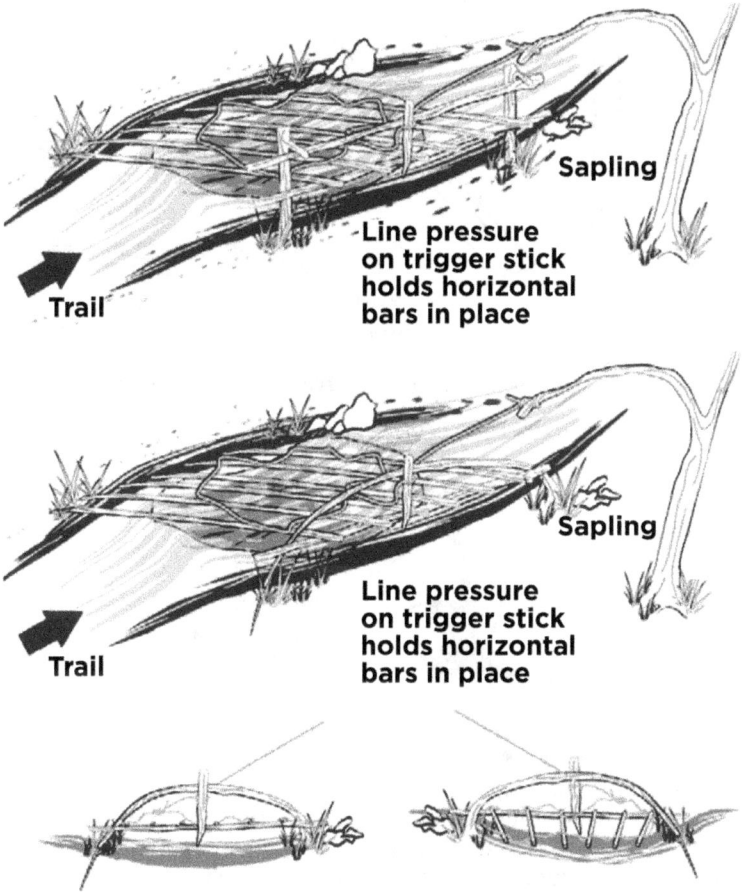

Sapling

**Line pressure
on trigger stick
holds horizontal
bars in place**

Trail

Sapling

**Line pressure
on trigger stick
holds horizontal
bars in place**

Trail

4-103. Deadfall. The deadfall is a trigger used to drop a weight onto a prey and crush it. The type of weight used may vary, but it should be heavy enough to kill or incapacitate the prey immediately. Construct the deadfall using three notched sticks. These notches hold the sticks together in a figure 4 pattern when under tension. Practice making this trigger beforehand; construction requires close tolerances and precise angles (see Figure 4-14 on page 140).

Figure 4-14. Deadfall

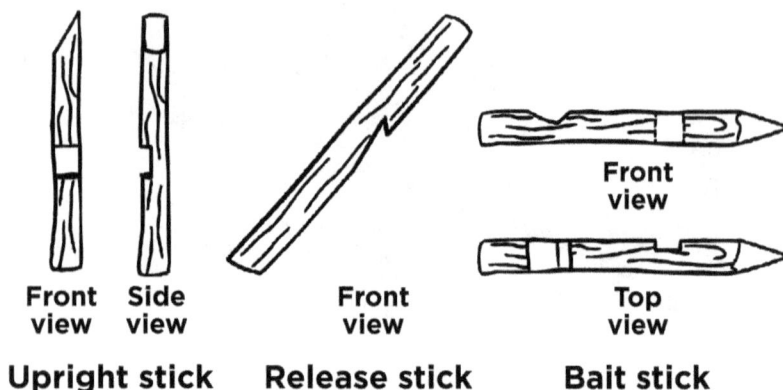

**Front   Side   Front   Top
view   view   view   view**

Upright stick   Release stick   Bait stick

4-104. Paiute[1] deadfall. The Paiute deadfall is similar to the standard (Figure 4-14) deadfall, but uses a piece of cordage and a catch stick. It has the advantage of being easier to set than the standard type. Tie one end of a piece of cordage to the lower end of the diagonal stick. Tie the other end of the cordage to another stick about 2 inches long. This stick is the catch stick. Bring the cord halfway around the vertical stick with the catch stick at a 90-degree angle. Place the bait stick with one end against the drop weight, or a peg driven into the ground, and the other end against the catch stick. When a prey animal disturbs the bait stick, it falls free, releasing the catch stick. As the diagonal stick flies up, the weight falls, crushing the prey. To increase the effectiveness of this trap, a small bait well may be dug into the bottom of the hole with bait placed in it to lure animals to the trap (see Figure 4-15 on page 141).

---

1. Pronounced "Pie-yoot" and named after the indigenous people of the Great Basin between the Rocky Mountains and the Sierra Nevada.

Figure 4-15. Paiute Deadfall

4-105. Bow trap. A bow trap is one of the deadliest traps. It is dangerous to man as well as animals. To construct this trap, build a bow and anchor it to the ground with pegs. Adjust the aiming point as the bow is anchored. Lash a toggle stick to the trigger stick. Two upright sticks driven into the ground hold the trigger stick in place at a point where the toggle stick engages the pulled bowstring. Place a catch stick between the toggle stick and a stake driven into the ground. Tie a tripwire or cordage to the catch stick and route it around stakes and across the game trail where it is tied off. When the prey trips the tripwire, the bow launches an arrow into it. A notch in the bow serves to help aim the arrow (see Figure 4-16 on page 142).

Note: Use caution—this is a lethal trap and should only be approached from the rear of the trap.

Figure 4-16. Bow Trap

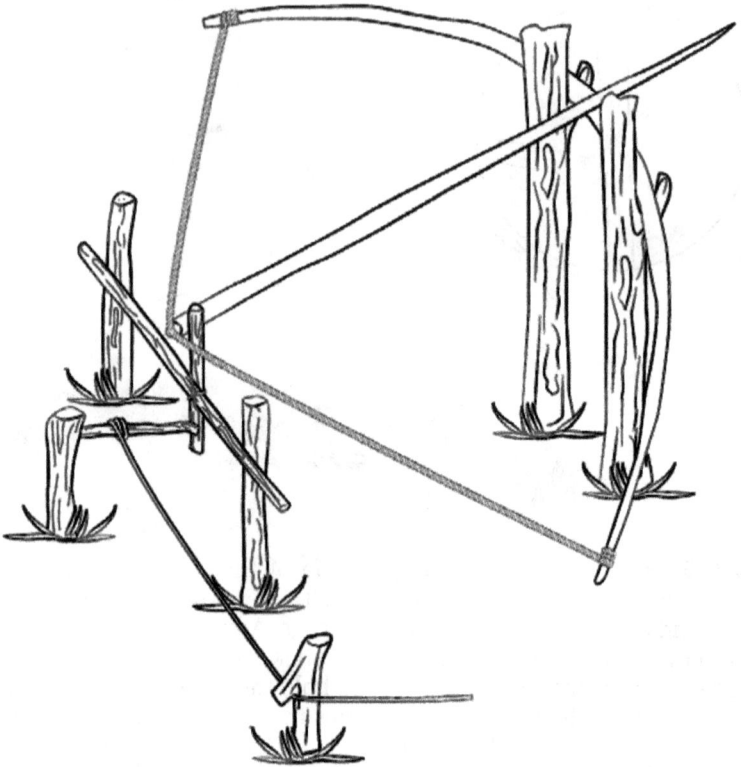

4-106. Pig spear shaft. To construct a pig spear shaft, select a stout pole about eight feet long. At the smaller end, firmly lash several small, sharp stakes. Lash the large end tightly to a tree along the game trail. Tie a length of cordage to another tree across the trail. Tie a sturdy, smooth stick to the other end of the cord. From the first tree, tie a tripwire or cord low to the ground, stretch it across the trail, and tie it to a catch stick. Make a slip ring from vines or other suitable material. Encircle the tripwire and the smooth stick with the slip ring. Emplace one end of another smooth stick within the slip ring and its other end against the second tree. Pull the smaller end of the spear shaft across the trail and position it between the short cord and the smooth stick.

4-107. As the animal trips the tripwire, the catch stick pulls the slip ring off the smooth sticks, releasing the spear shaft that springs across the trail and impales the prey against the tree. (see Figure 4-17 on page 143).

Note: Use caution—this is a lethal trap and should only be approached from the rear of the trap.

Figure 4-17. Pig Spear Shaft

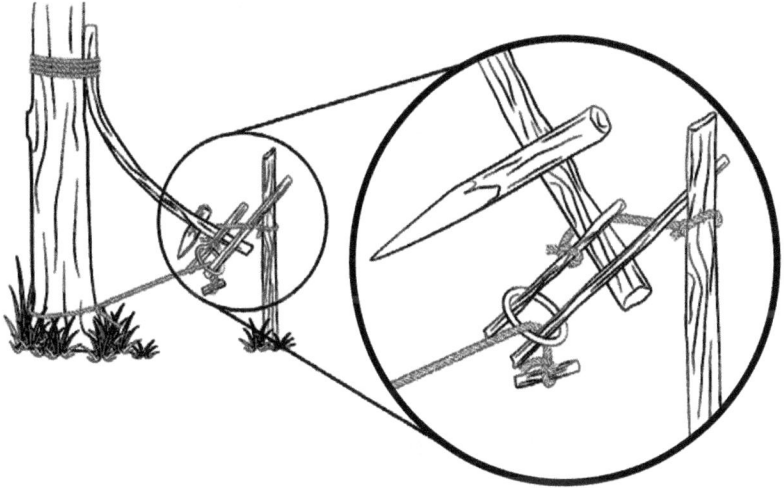

4-108. Bottle trap. A bottle trap is a simple trap for mice and moles. Dig a hole 12~18 inches deep that is wider at the bottom than at the top. Make the top of the hole as small as possible. Place a piece of bark or wood over the hole with small stones under it to hold it up 1 to 2 inches off the ground. Mice or moles will hide under the cover to escape danger and fall into the hole. They cannot climb out because of the wall's backward slope (see Figure 4-18 on page 144).

Note: Note. Use caution when checking this trap—it is an excellent hiding-place for snakes.

Figure 4-18. Bottle Trap

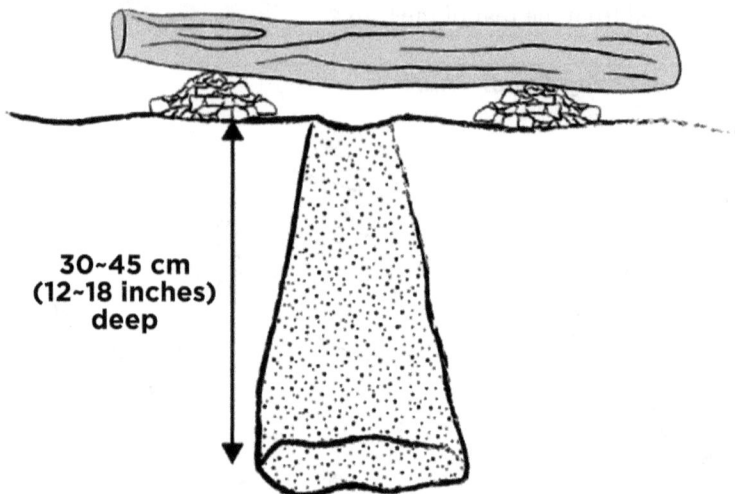

**30~45 cm
(12~18 inches)
deep**

KILLING DEVICES

4-109. The rabbit stick, the spear, the bow and arrow, and the sling are several types of killing devices constructed to capture small game to aid in survival.

RABBIT STICK

4-110. One of the simplest and most effective killing devices is a stout stick as long as an arm, from fingertip to shoulder, called a "rabbit stick." Thrown overhand or sidearm and with considerable force, it is best thrown so that it flies sideways, increasing the chance of hitting the target. It is very effective against small game that stops and freezes as a defense.

SPEAR

4-111. Improvise a spear by sharpening the end of a long pole. The spear will be useful to kill small game and to fish. Jab with the spear—do not throw it.

SLING

4-112. A sling can be made by tying two pieces of cordage, each about twenty-four inches long, at opposite ends of a palm-sized piece of leather or cloth. Place a rock in the cloth, wrap one cord around the middle finger, and hold it in the palm. Hold the other cord between the forefingers and thumb. To throw the rock, spin the sling several times in a circle and release the cord between the thumb and forefinger. The sling is very effective against small game, but proficiency with the sling requires practice.

BUTCHERING MAMMALS

4-113. Once the animal is skinned, break it down into usable-size pieces. This will allow more options for transporting, cooking, and preserving. Butchering involves removing the muscle tissue from the skeletal structure of the animal. Trim off the connective tissue, trim off the fat (save for consumption), and cut the flesh into smaller pieces similar to what would be seen in a butcher shop. The meat can be cut into steaks, roasts, stew meat cubes, and thin slices for drying and smoking.

SKINNING MAMMALS

4-114. Once a mammal has been procured, it will need to be skinned. There are two general ways to skin a mammal—the big game method and the glove skinning method.

Big Game

4-115. When the mammal is a large animal, the big game skinning method should be used. The first step in skinning is to turn the animal on its back and, with a sharp knife, cut through the skin on a straight line from the tail bone to a point under its neck as illustrated in Figure 4-19 on page 147. In making this cut, pass around the anus and, with great care, press the skin open until the first two fingers can be inserted between the skin and the thin membrane enclosing the entrails. When the fingers can be forced forward, place the blade of the knife between the fingers, blade up, with the knife held firmly. While forcing the fingers forward, palm upward, follow with the knife blade, cutting the skin but not cutting the membrane.

4-116. If the animal is a male, cut the skin parallel to, but not touching, the penis. If the tube leading from the bladder is accidentally cut, a messy job and unclean meat will result. If the gall or urine bladders are broken, washing will help clean the meat. Otherwise, it is best not to wash the meat but to allow it to form a protective glaze. On reaching the ribs, it is no longer possible to force the fingers forward, because the skin adheres more strongly to flesh and bone. Furthermore, care is no longer necessary. The cut to point C (in Figure 4-19 on page 147) can be quickly completed by alternately forcing the knife under the skin and lifting it.

4-117. With the central cut completed, make side cuts consisting of incisions through the skin, running from the central cut up the inside of each leg to the knee and hock joints. Make cuts around the front legs just above the knees and around the hind legs above the hocks. Make the final cross cut at point C (in Figure 4-19 on page 147), and then cut completely around the neck and in back of the ears. Once the cuts are complete, the skinning process can begin. After skinning down the animal's side as far as possible, roll the carcass on its side to skin the back. Then spread out the loose skin to prevent the meat from touching the ground and turn the animal on the skinned side. Follow the same procedure on the opposite side until the skin is free. In opening the membrane which encloses the entrails, follow the same procedure used in cutting the skin by using the fingers of one hand as a guard for the knife and separating the intestines from the membrane. This thin membrane along the ribs and sides can be cut away for a better view.

4-118. Be careful to avoid cutting the intestines or bladder. The large intestine passes through an aperture in the pelvis. Using a knife, separate this tube from the bone surrounding it. Tie a knot in the bladder tube to prevent the escape of urine. With these steps complete, the entrails can be easily disengaged from the back and removed from the carcass. After gutting is completed, it may be advisable to hang the animal.

Figure 4-19. Deer Skinning Example

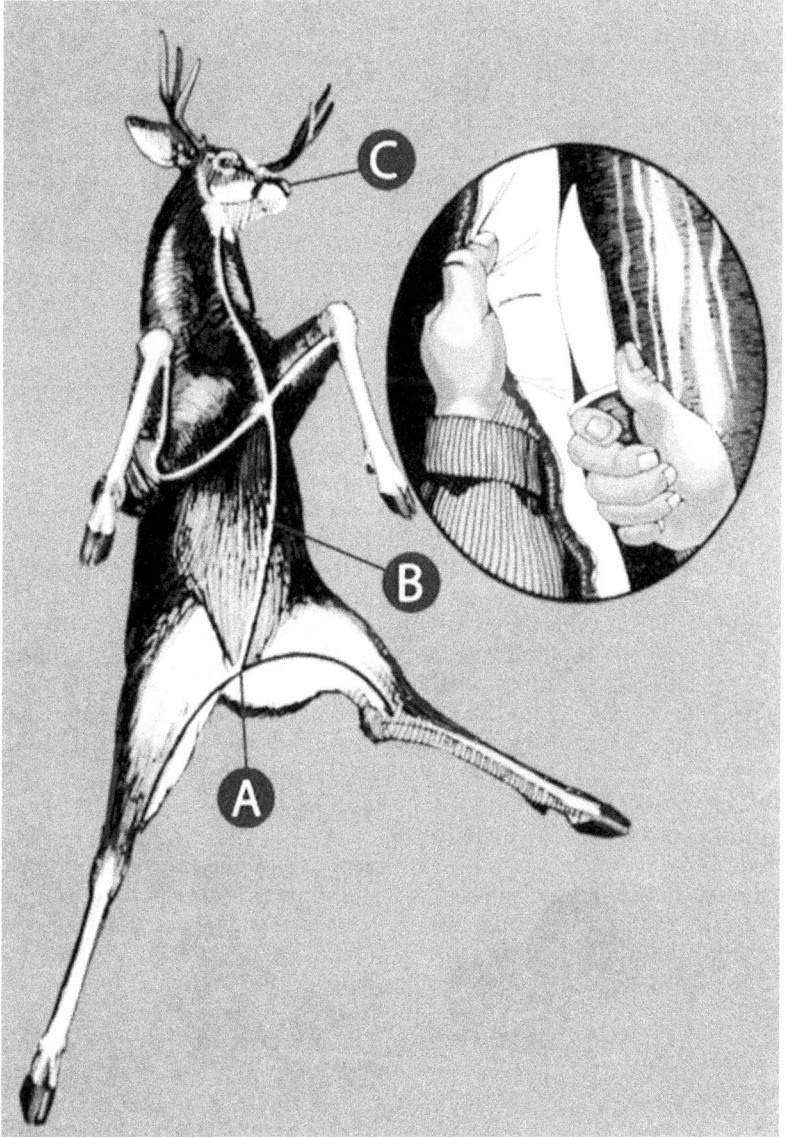

Note: If the animal is hot, gut the animal before skinning it.

4-119. The intestines of a well-conditioned animal are covered with a lace-like layer of fat which can be lifted off and placed on nearby bushes to dry for later use. The gall bladder which is attached to the liver of some animals should be carefully removed. If it ruptures, the bile will taint anything it touches. Clean the knife if necessary. The kidneys are imbedded in the back, forward of the pelvis, and are covered with fat. Running forward from the kidneys on each side of the backbone are two long strips of chop meat or muscle called tenderloin and/or back strap. Eat this after the liver, heart, and kidneys as it is usually very tender. Edible meat can also be removed from the head, brisket, ribs, backbone, and pelvis.

4-120. Large animals should be quartered. To do this, cut down between the first and second rib and then sever the backbone with an axe or machete. Cut through the brisket of the front half and then chop lengthwise through the backbone to produce the front quarters. On the rear half, cut through the pelvic bone and lengthwise through the backbone. To make the load lighter and easier to transport, the bones can be removed with a knife. Butchering is the final step and is simplified for survival purposes. The main purpose is to cut the meat into manageable-sized portions.

Small Game

4-121. Glove skinning is performed on small game. The initial cuts are made down the insides of the back legs. The skin is then peeled back so that the hindquarters are bare and the tail is severed. To remove the remaining skin, pull it down over the body in much the same manner a pullover sweater is removed. Sever the head and front feet to remove the skin from the body. To remove the internal organs, make a cut into the abdominal cavity without puncturing the organs. This cut must run from the anus to the neck. There are muscles that connect the internal organs to the trunk, and they must be severed to allow the viscera to be removed.

4-122. If the isolated person does not have a knife and only access to rudimentary tools like a sharp rock or dull piece of metal, there is a method of skinning sometimes referred to as the "pants and shirt" method of skinning. This method is simple and works very well on small to medium size game. The isolated person will separate the skin around the midline of the animal by making an incision around the animal with a cutting implement or by tearing it with their hands. The isolated person will then pull the top and bottom pieces of skin in the opposite direction until they are fully removed from the carcass. This method is very fast and can be completed from a concealed position when needed.

BIRDS

4-123. Birds provide a concentrated form of protein and amino acids. Isolated persons eating birds will benefit from their bodies being able to obtain energy and repair themselves from the protein. As with all wild animals, understanding the birds' habits will offer a better chance of catching them. Some species of bird can be captured from their roost at night by hand. During the nesting season, some species will not leave the nest even when approached.

4-124. Birds tend to have regular flyways going from the roost to a feeding area, to water, and so forth. Careful observation should reveal where these flyways are and indicate good areas for catching birds in nets stretched across the flyways (see Figure 4-20). A small gill net on a wooden frame with a disjointed stick for a trigger can also be used. Birds can be caught on baited fishhooks or with simple loop snares. Roosting sites and waterholes are some of the most promising areas for trapping and snaring.

Figure 4-20. Mist Net Bird Snare

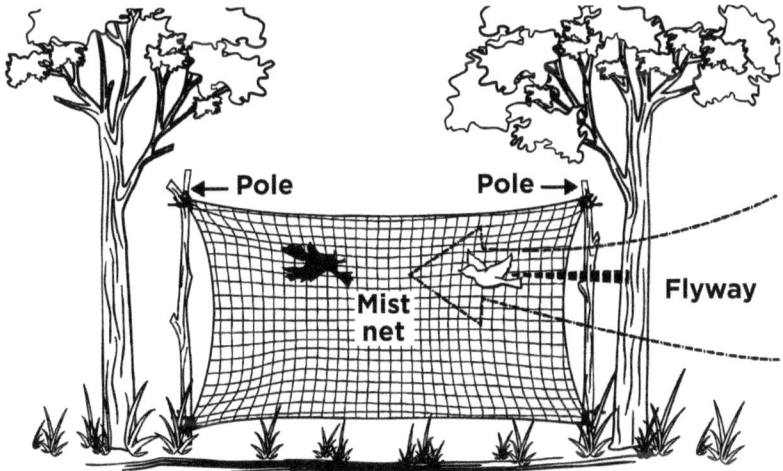

4-125. An Ojibwa bird pole is a snare that has been used by Native Americans for centuries. To be effective, it should be placed in a relatively open area away from tall trees. For best results, pick a spot near feeding areas, dusting areas, or watering holes.

4-126. Cut a pole 6 to 7 feet long and trim away all limbs and foliage. Do not use resinous wood such as pine. Sharpen the upper end to a point, and then drill a small-diameter hole 2~3 inches down from the top. Cut a small stick 4~6 inches long and shape one end so that it will almost fit into the hole. This is the perch. Plant the long pole in the ground with the pointed end up. Tie a small weight, about equal to the weight of the targeted species, to a length of cordage. Pass the free end of the cordage through the hole and tie a slip noose that covers the perch. Tie a single overhand knot in the cordage and place the perch against the hole. Allow the cordage to slip through the hole until the overhand knot rests against the pole and the top of the perch. The knot must be smaller than the hole, enabling the knot to slip through the hole.

4-127. The tension of the overhand knot against the pole and perch will hold the perch in position. Spread the noose over the perch, ensuring that it covers the perch and drapes over on both sides. Most birds prefer to rest on something above ground and will land on the perch. As soon as the bird lands, the perch will fall, releasing the overhand knot and allowing the weight to drop. The noose will tighten around the bird's feet, capturing it. Ensure the weight is not too heavy, otherwise it will cut off the bird's feet, allowing it to escape. Another variation would be to use spring tension such as a tree branch in place of the weight (see Figure 4-21).

Figure 4-21. Ojibwa Bird Pole

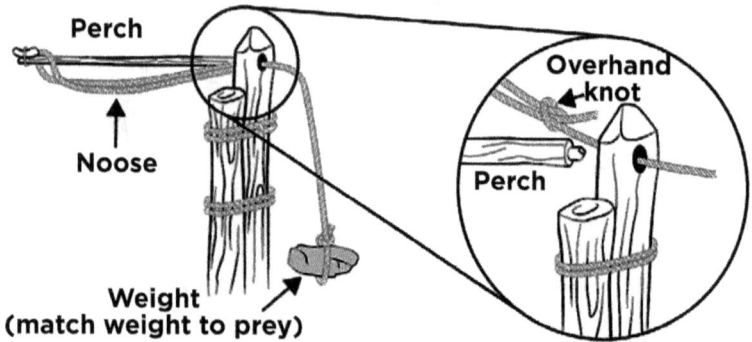

4-128. Butchering birds is done by first plucking the bird to remove its feathers. The bird can be placed in a container of boiling water to loosen the feathers and make it easier to pluck. The bird can also be skinned to remove the feathers, but skinning removes some of the bird's nutritional value. After plucking or skinning the bird, open up the body cavity and remove the entrails, saving the heart and liver. Most of the usable meat from a bird comes from the breast.

4-129. For larger birds, the legs and thighs can produce a fair amount of meat. Next, cut off the feet. With a knife or improvised tool, cut in between the joints of the bird to separate the skeletal structure. Because of the fragility of a bird's muscle and skeletal structure most of it can be processed with just the hands, requiring no tools.

4-130. Make sure to wash your hands and any tools after working with raw bird meat to avoid the transfer of bacteria. It is best to leave the meat on the bone for preparation. Most birds except sea birds should be cooked with their skin left on. Sea birds will taste better if the oily skin is removed before cooking. Carrion-eating birds, such as vultures, must be boiled for at least 20 minutes to kill parasites before further cooking and eating. Cooking techniques are the same as with other game animals.

4-131. Bird meat can be boiled, baked, roasted, or fried. Bird meat should be fully cooked to eliminate any chance of contracting food-borne illnesses. Preservation methods are the same as other meat. If in or near the ocean, the meat can be placed in a saltwater bath to preserve it (wash the salt off the meat before cooking).

INSECTS

4-132. The most abundant and easily-caught life forms on earth are insects. Many insects provide 65~80 percent protein compared to only 20 percent for beef. This makes insects an important source of nutrition for the isolated person. The problem with insects is that they have to be gathered in large numbers to get the same amount of protein you would receive from a rabbit or other piece of meat.

4-133. Insects to avoid include all adults that bite or sting, and hairy or brightly colored insects. Hairy caterpillars and insects that have a pungent odor should also be avoided as well as spiders and common disease carriers such as ticks, flies, and mosquitoes.

4-134. Ants such as the white ant, the carpenter ant, and the honey ant are edible and were a staple in many native cultures. Termites, bee larvae, locusts, dragonflies, bumblebees, cockroaches, locust grasshoppers, golden June beetles, crickets, wasp larvae, and silkworm larvae are all used for food. Flying insects can be attracted to a light source. If you place a piece of material in front of the light source with a vessel of water below it, the flying insects will fly into the material and some will fall into the water, remaining for collection and consumption.

4-135. Rotting logs lying on the ground are excellent places to look for a variety of insects including ants, termites, beetles, and grubs. Also, do not overlook nests on or in the ground. Grassy areas, such as fields, are good areas to search because the insects are easily seen. Stones, boards, or other materials lying on the ground provide insects with good nesting sites.

4-136. Preparation of insects involves removing any stinging apparatus from the insect before consuming. Also, remove their legs, wings, and heads before eating them. Insects with hard outer shells will have parasites and should be fully cooked before consumption. Insects can be cooked using all available cooking methods. Many insects can be preserved for long periods by keeping them alive. Other forms of food preservation are appropriate as well.

4-137. Insects can be cooked by roasting, baking, frying, and boiling. They can be added to other foods such as plants and other meat, making a stew. This will provide needed protein while avoiding any potential aversion to eating insects. Earthworms are high in protein and can be found in humus soil and in the root ball of grass clumps, or watch for them after rain on the ground's surface. They should be prepared as follows: soak them in a bath of clear, potable water for a day, squeeze all of the guts to eliminate waste products, and clean them. Then cook them by roasting, baking, frying, or boiling.

REPTILES

4-138. Reptiles are a good protein source and relatively easy to catch. Reptiles can be found moving across open ground. They can be procured using a long, forked stick. Place the stick behind the reptile's head to immobilize its ability to strike and inject you with venom. Then cut off the head at least 1~2 inches (4~6 inches or more for larger specimens) behind the head to prevent accidental envenomation from the venom sacs left in the snake's head. Snakes have been known to retain a bite reflex for up to 24 hours after the head is removed. Bury the head to prevent it from accidentally being stepped on at a later time. Cut up the belly of the snake and remove the innards. Peel the skin back. Pull the skin in one hand and the body of the snake in the other until the skin is removed (see Figure 4-22). Traps with bait may also prove useful in gathering reptiles.

Figure 4-22. Cleaning a Snake

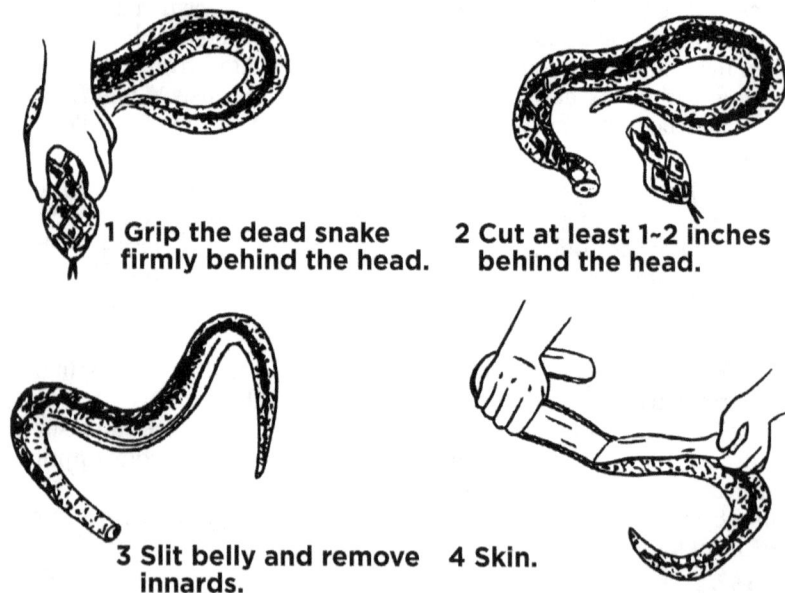

1 Grip the dead snake firmly behind the head.

2 Cut at least 1~2 inches behind the head.

3 Slit belly and remove innards.

4 Skin.

4-139. Thorough cooking and hand washing is imperative with reptiles. All reptiles are considered carriers of salmonella, which exists naturally on their skin. Turtles and snakes are especially known to infect man. Salmonella can be deadly to an isolated person in an undernourished state, with a weak immune system. Cook food thoroughly and be especially fastidious in washing hands after handling any reptile.

4-140. Lizards are plentiful in most parts of the world. They have dry, scaly skin and five toes on each foot. The only venomous ones are the Gila monster and the Mexican bearded lizard. Care must be taken when handling and preparing the iguana and the monitor lizard, as they commonly harbor the salmonella bacteria in their mouth and teeth. The tail meat is the best tasting and easiest to prepare.

4-141. Turtles are a very good source of meat. There are actually seven different flavors of meat in each snapping turtle. Most turtle meat comes from the front and rear shoulder area, although a large turtle may have some on its neck. The box turtle (which resembles a tortoise) is a commonly encountered turtle that should not be eaten. It feeds on poisonous mushrooms and may build up a highly toxic poison in its flesh. Cooking does not destroy this toxin. Also avoid the hawksbill turtle found in the Atlantic Ocean, because of its poisonous thorax gland (see Figure 4-23).

Figure 4-23. Turtles with Poisonous Flesh

Box turtle   **Hawksbill turtle**

4-142. Cook reptiles in the same manner as other small game using all cooking methods available. Isolated persons should concentrate on eating the fleshy parts of the reptile. For example, a turtle has meat near its legs and where the body meets the fins, neck, and tail. These areas can be fried or cooked in a stew. All meat should be thoroughly cooked. Reptile eggs can also be eaten. Reptiles can be kept alive to preserve them, or use other methods of preservation previously discussed to keep them fresh for consumption.

AMPHIBIANS

4-143. Frogs are easily found around bodies of fresh water. Frogs seldom move from the safety of the water's edge. At the first sign of danger, they plunge into the water and bury themselves in the mud and debris. Frogs are characterized by smooth, moist skin. There are few poisonous species of frogs. Avoid any brightly colored frog or one that has a distinct "X" mark on its back as well as all tree frogs. Do not confuse toads with frogs. Toads may be recognized by their dry, "warty" or bumpy skin. They are usually found on land in drier environments.

4-144. Several species of toads secrete a poisonous substance through their skin as a defense against attack. Therefore, to avoid poisoning, do not handle or eat toads. Do not eat salamanders; only about 25 percent of all salamanders are edible, so it is not worth the risk of possibly selecting a poisonous variety. Salamanders are found around the water. They are characterized by smooth, moist skin and have only four toes on each foot. Amphibians provide a good source of protein (see Figure 4-24).

4-145. A frog gig, which is a long spear with several tines resembling a pitchfork on the end, can be used to spear an amphibian at the water's edge. Most of the meat content from frogs will come from their legs. Amphibians should be cooked like any other small game; roasting on a stick works well. The meat should be thoroughly cooked. Preservation of amphibians as food is the same as stated for other meat.

Figure 4-24. Amphibians

Bullfrog

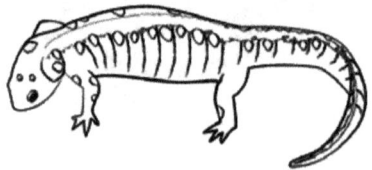
Spotted salamander

Toad

Newt

FISH

4-146. Fish represent a good source of protein and fat. They offer some distinct advantage in that they are usually more abundant than mammalian wildlife, and the methods to procure them are mostly silent. To be successful at catching fish, their habits must be known. For instance, fish tend to feed heavily before a storm. Fish are not likely to feed after a storm when the water is muddy and swollen. Light often attracts fish at night.

4-147. When there is a heavy current fish rest in places where there is an eddy, such as near rocks. Fish also gather where there are deep pools, under overhanging brush, and in and around submerged foliage, logs, or other objects that offer them shelter. There are no poisonous freshwater fish. However, the catfish species has sharp, needlelike protrusions on its dorsal fins and barbels. These can inflict painful puncture wounds that quickly become infected.

4-148. Cook all fish to kill parasites, particularly freshwater fish. Most wild-caught fish contain parasites such as nematodes (roundworms), tapeworms and flukes, particularly in their internal organs.

4-149. Most fish encountered are edible. However, the organs of some species are always poisonous to man; other fish can become toxic because of elements in their diets. Ciguatera is a form of human poisoning caused by the consumption of subtropical and tropical marine fish, which have accumulated naturally occurring toxins through their diet. These toxins build up in the fish's tissues. The toxins are known to originate from several algae species that are common to ciguatera-endemic regions in the lower latitudes. Cooking does not eliminate the toxins; neither does drying, smoking, or marinating.

4-150. Marine fish most commonly implicated in ciguatera poisoning include barracudas, jacks, mackerel, triggerfish, snappers, and groupers. Many other species of warm water fishes harbor ciguatera toxins. The occurrence of toxic fish is sporadic and not all fish of a given species or from a given locality will be toxic. Other examples of poisonous saltwater fish are the porcupine fish, the cowfish, the thorn fish, the oil fish, and the puffer fish (see Figure 4-25 on page 156).

Figure 4-25. Fish with Poisonous Flesh

Cowfish
(15~30 cm, 6~12 inches)

Oilfish
(90~150 cm, 36~60 inches)

Red snapper
(60~90 cm, 24~36 inches)

Jack
(about 60 cm, 24 inches)

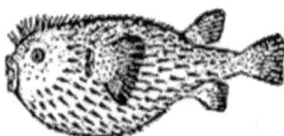

Porcupine fish
(about 30 cm, 12 inches)

Trigger fish
(30~60 cm, 12~24 inches)

Puffer
(25~38 cm, 10~15 inches)

Thorn fish
(about 30 cm, 12 inches)

FISHING DEVICES

4-151. Isolated persons can make fishhooks, nets, and traps. Field-expedient fishhooks can be made from pins, needles, wire, small nails, or any piece of metal (see Figure 4-26 on page 157). Wood, bone, coconut shells, thorns, flint, seashells, tortoise shells, or a combination of these items can also be used.

Figure 4-26. Improvised Fish Hooks

**Carved wood
gorge hook**

Wire

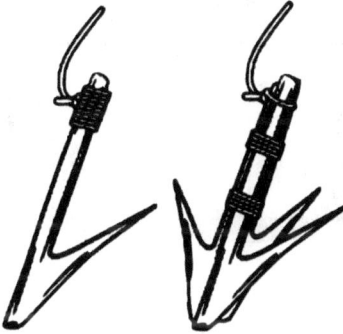

Thorn hooks

**Carved
wood shanks**

Wooden Hook

4-152. To make a wooden hook, cut a piece of hardwood about 1 inch long and about ¼ inch in diameter to form the shank. Cut a notch in one end in which to place the point. Place the point (a piece of bone, wire, or nail) in the notch. Hold the point in the notch and tie it securely so that it does not move out of position. This is a large hook; to make smaller hooks, use smaller material.

Gorge or Skewer

4-153. A gorge or skewer is a small shaft of wood, bone, metal, or other material. It is sharp on both ends and notched in the middle where cordage is tied. Bait the gorge by placing a piece of bait on it lengthwise. When the fish swallows the bait, it also swallows the gorge. If you are tending the fishing line when the fish bites, you should not attempt to pull on the line to set the hook as you would with a conventional hook. Allow the fish to swallow the bait to get the gorge as far down its throat as possible before the gorge sets itself.

Stakeout

4-154. A stakeout is a fishing device that can be used covertly in a non-permissive (hostile or high-threat) environment. To construct a stakeout, drive two supple saplings into the bottom of the lake, pond, or stream with their tops just below the water's surface. Tie a cord between them just slightly below the surface. Tie two short cords with hooks or gorges to this cord, ensuring that they cannot wrap around the poles or each other. They should also not slip along the long cord. Bait the hooks or gorges (see Figure 4-27).

Figure 4-27. Stakeout

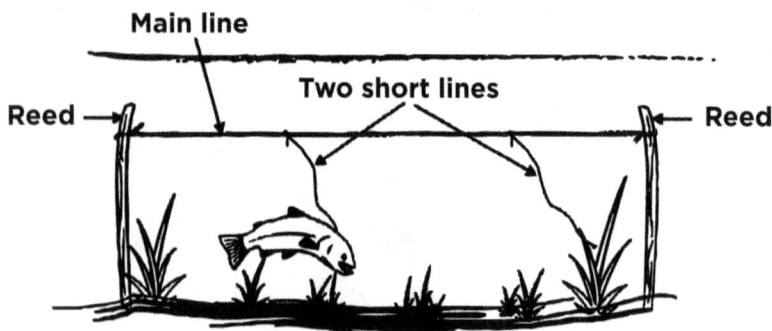

Gill Net

4-155. If a gill net is not available, isolated persons can make one using parachute suspension line or similar material (see Figure 4-28 on page 159). Remove the core lines from the suspension line and tie the casing between two trees. Attach several core lines to the casing by doubling them over and tying them with Prusik knots or girth hitches. These lines should be six times the desired depth of the net (for example, a 6-foot piece of string girth-hitched over the casing will yield two 3-foot pieces, which after completing the net, will provide a 1-foot deep net).

4-156. The length of the desired net and the size of the mesh determine the number of core lines used and the space between them. The recommended size of the spaces in the net mesh is about 1 inch square. Starting at one end of the casing, tie the second and the third core lines together using an overhand knot. Then tie the fourth and fifth, sixth and seventh, and so on, until the last core line is reached. All core lines should now be tied in pairs with a single core line hanging at each end. Start the second row with the first core line, tie it to the second, the third to the fourth, and so on.

Figure 4-28. Making a Gill Net

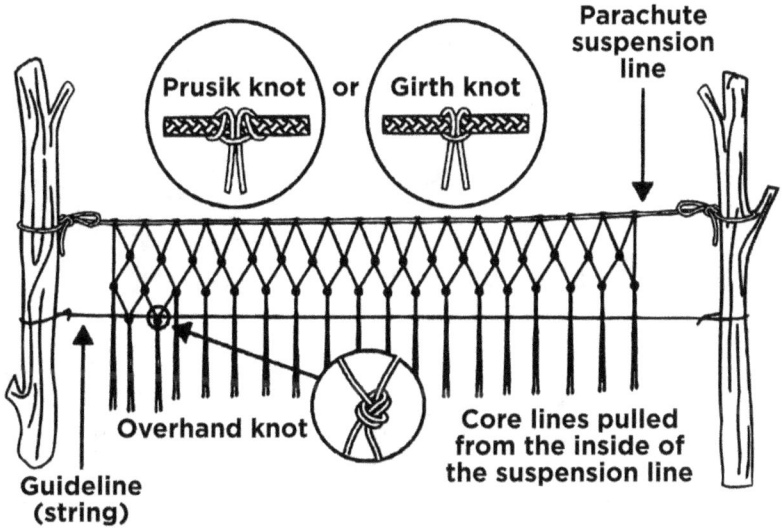

4-157. To keep the rows even and to regulate the size of the mesh, tie a guideline to the trees. Position the guideline on the opposite side of the net you are working on. Move the guideline down after completing each row. The lines will always hang in pairs and you should always tie a cord from one pair to a cord from an adjoining pair. Continue tying rows until the net is the desired width. Thread a suspension line casing along the bottom of the net to strengthen it. Use the gill net as shown in Figure 4-29 on page 160. Angling the gill net will help to reduce the amount of debris that may accumulate in the net. Check it frequently.

Figure 4-29. Setting a Gill Net in a Stream

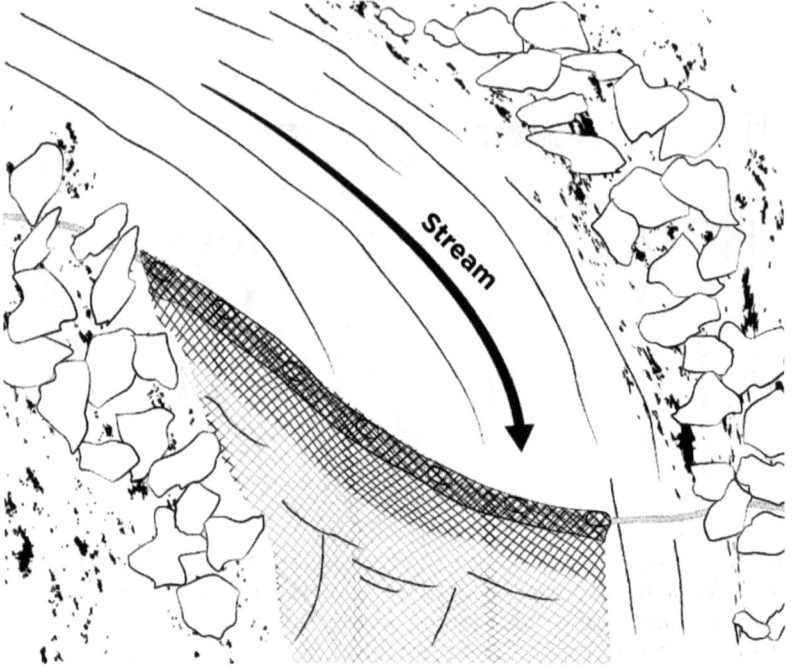

Fish Traps

4-158. There are several methods available for trapping fish (see Figure 4-30 on page 161). One such method is the fish basket. Fish baskets are constructed by lashing several sticks together with vines into a funnel shape. The top is closed, leaving a hole large enough for the fish to swim through.

Figure 4-30. Various Types of Fish Traps

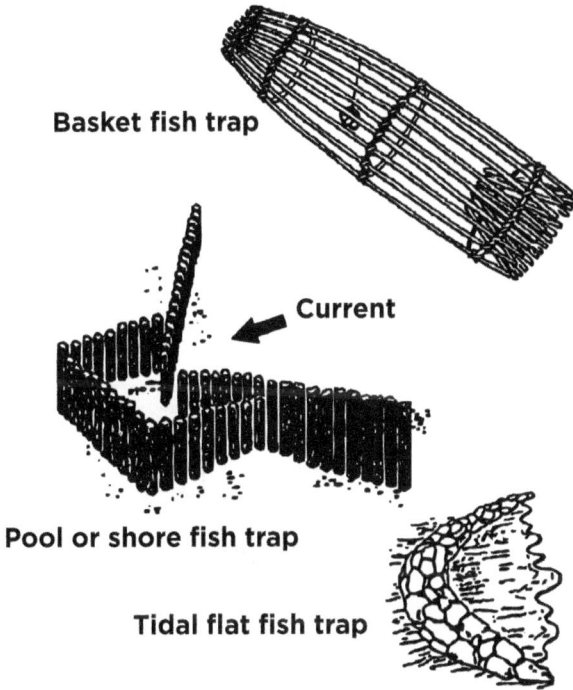

Basket fish trap

Current

Pool or shore fish trap

Tidal flat fish trap

4-159. Fish traps can also be used to catch saltwater fish, as schools regularly approach the shore with the incoming tide and often move parallel to the shore. Pick a location at high tide and build the trap at low tide. On rocky shores, use natural rock pools. On coral islands, use natural pools on the surface of reefs by blocking the openings as the tide recedes. On sandy shores, use sandbars and the ditches they enclose. Build the trap as a low stone wall extending outward into the water and forming an angle with the shore.

4-160. Near shallow water (about waist deep) where fish are large and plentiful, they can be speared. To make a spear, cut a long, straight sapling. Sharpen the end to a point or attach a knife, a jagged piece of bone, or sharpened metal. A spear can also be made by splitting the shaft a few inches down from the end and inserting a piece of wood to act as a spreader, then sharpening the two separated halves to points (see Figure 4-31 on page 162).

4-161. To spear fish, find an area either where fish gather or where there is a fish run. Place the spear point into the water and slowly move it toward the fish. Then, with a sudden push, impale the fish on the stream bottom. Do not try to lift the fish with the spear, as it will probably slip off; hold the spear with one hand and grab and hold the fish with the other. Do not throw the spear, especially if the point is a knife. An isolated person cannot afford to lose a knife in a survival situation. Be alert to the problems caused by light refraction when looking at objects in the water: aim lower than the object, usually at the bottom of the fish.

Figure 4-31. Types of Spear Points

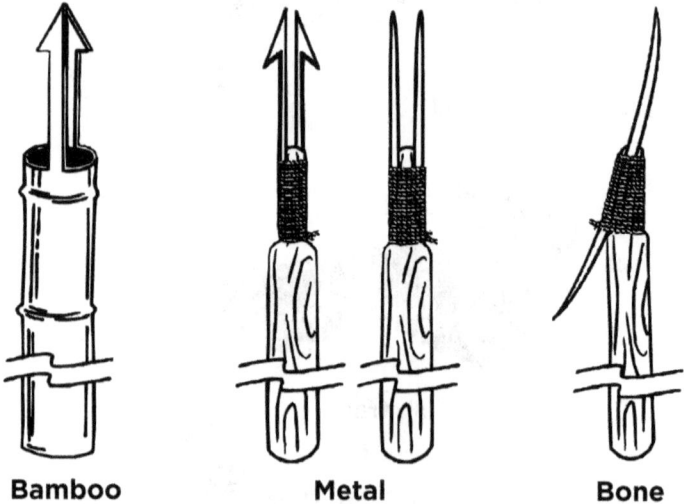

Bamboo   **Metal**   **Bone**

Chop Fishing

4-162. Chop fishing is also an option if you are equipped with a machete or similar instrument. At night, in an area with high fish density, use a light to attract fish. Then, you can gather fish by using the backside of the machete blade to strike them. Do not use the sharp side of the weapon as it will cut the fish in two pieces and one piece of the fish may be lost.

Poison

4-163. Poison may also be used to catch fish. Poison works quickly and allows an isolated person to remain concealed while it takes effect. It also enables you to catch several fish at one time. When using fish poison, be sure to gather all of the affected fish, because many dead fish floating downstream could arouse suspicion. Some plants that grow in warm regions of the world contain rotenone, a substance that stuns or kills cold-blooded animals but does not harm persons who eat the animals. The best place to use rotenone, or rotenone-producing plants, is in ponds or the headwaters of small streams containing fish. Rotenone works quickly on fish in water 70° F (21° C) or above. The fish rise helplessly to the surface. It works slowly in water 50 to 70° F (10 to 21° C) and is ineffective in water below 50° F (10° C). The following plants and materials, used as indicated, will stun or kill fish (see Figure 4-32 on page 164):

- **Anamirta cocculus**. This woody vine grows in southern Asia and on islands of the South Pacific. Crush the bean-shaped seeds and throw them in the water.

- **Croton tiglium**. This shrub or small tree grows in waste areas on islands of the South Pacific. It bears seeds in three angled capsules. Crush the seeds and throw them into the water.

- **Barringtonia**. These large trees grow near the sea in Malaya and parts of Polynesia. They bear a fleshy one-seeded fruit. Crush the seeds, bark, and throw them into the water.

- **Derris eliptica**. This large genus of tropical shrubs and woody vines is the main source of commercially produced rotenone. Grind the roots into a powder and mix with water. Throw a large quantity of the mixture into the water.

- **Duboisia**. This shrub grows in Australia and bears white clusters of flowers and berrylike fruit. Crush the plants and throw them into the water.

- **Tephrosia**. This species of small shrubs, which bears bean-like pods, grows throughout the tropics. Crush or bruise bundles of leaves and stems and throw them into the water.

- **Lime** (the mineral material, not the fruit). Lime can be acquired from commercial sources and in agricultural areas that use large quantities of it. It can also be produced by burning coral or seashells. Throw the lime into the water.

- **Nut husks**. Crush green husks from butternuts or black walnuts. Throw the husks into the water.

Figure 4-32. Fish-poisoning Plants

Anamirta cocculus

Croton tiglium

Barringtonia

Derris elliptica

Duboisia

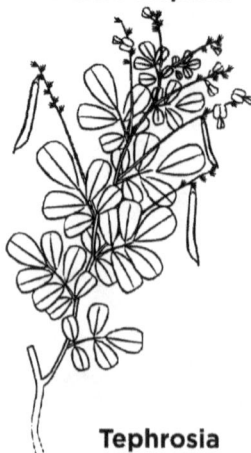

Tephrosia

PREPARING FISH FOR CONSUMPTION

4-164. You must know how to prepare fish and game for cooking and storage in a survival situation. Improper cleaning or storage can result in inedible fish or game. Do not eat fish that appears spoiled. Cooking does not ensure that spoiled fish will be edible. Signs of spoilage include—

- Sunken eyes.
- Peculiar odor.
- Suspicious color. (Gills should be red to pink; scales should be a pronounced shade of gray, not faded.)
- Dents that stay in the fish's flesh after pressed with a thumb.
- Slimy, rather than moist or wet, body.
- Sharp or peppery taste.

4-165. Eating spoiled or rotten fish may cause diarrhea, nausea, cramps, vomiting, itching, paralysis, or a metallic taste in the mouth. These symptoms appear suddenly, one to six hours after eating. Induce vomiting if symptoms appear.

4-166. Fish spoils quickly after death, especially on a hot day. Prepare fish for eating as soon as possible after catching it. Cut out the gills and the large blood vessels that lie near the spine. Gut fish that are more than 4 inches long. Scale or skin the fish.

4-167. A whole fish can be impaled on a stick and cooked over an open fire. However, boiling the fish with the skin on is the best way to get the most food value. The fats and oil are under the skin and, by boiling, the juices can be saved for broth. Any of the methods used to cook plant food may also be used to cook fish. Pack fish into a ball of clay and bury it in the coals of a fire until the clay hardens. Break open the clay ball to get to the cooked fish. Fish is done when the meat flakes off. If fish is to be kept for later consumption, smoke or fry it. To prepare fish for smoking, cut off the head and remove the backbone.

MOLLUSKS

4-168. The mollusk class includes octopuses and freshwater and saltwater shellfish such as snails, clams, mussels, bivalves, barnacles, periwinkles, chitons, and sea urchins (see Figure 4-33 on page 166). Bivalves, similar to our freshwater mussel and terrestrial and aquatic snails, may be found worldwide under all water conditions. River snails or periwinkles are plentiful in rivers, streams, and lakes of northern coniferous forests. These snails may be pencil-point or globular in shape. In fresh water, look for mollusks in the shallows, especially in water with a sandy or muddy bottom. Look for the narrow trails they leave in the mud or for the dark elliptical slit of their open valves. Near the sea, look in the tidal pools and the wet sand. Rocks along beaches or extending as reefs into deeper water often bear clinging shellfish. Snails and limpets cling to rocks and seaweed from the low water mark upward. Large snails, called chitons, adhere tightly to rocks above the surf line. Snails are a good source of calcium, magnesium and vitamin C. Ensure snails are cooked thoroughly. Mussels usually form dense colonies in rock pools, on logs, or at the base of boulders. Steam, boil or bake mollusks in the shell. They make excellent stews in combination with greens and tubers.

Figure 4-33. Edible Mollusks

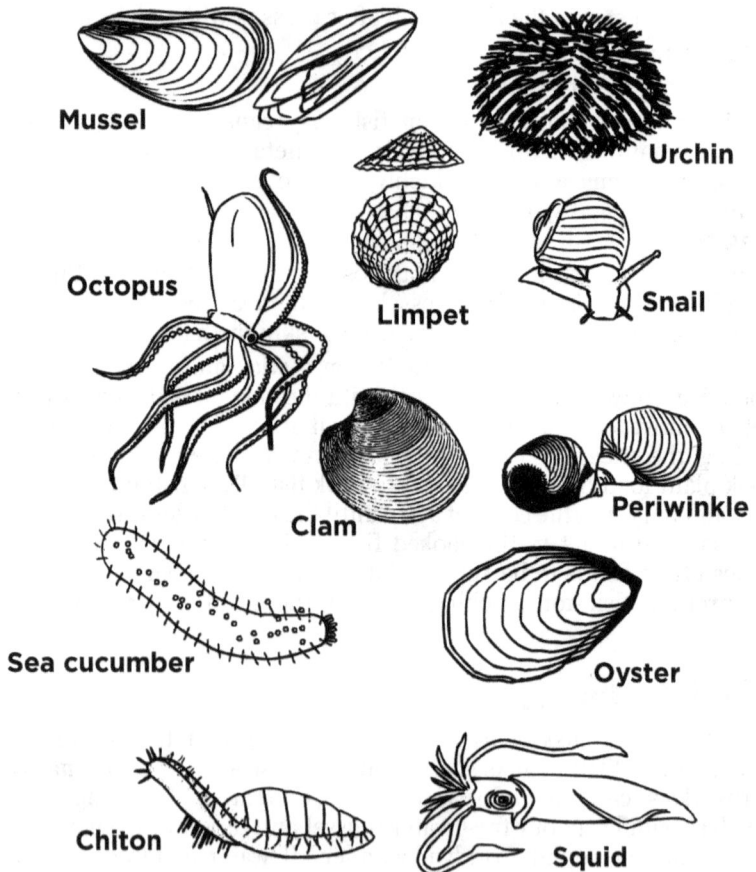

Mussel

Urchin

Octopus

Limpet

Snail

Clam

Periwinkle

Sea cucumber

Oyster

Chiton

Squid

CAUTION: Mussels may be poisonous in tropical areas during the summers. If a noticeable red tide (coastal water discoloration due to algal bloom) has occurred within 72 hours, do not eat any fish or shellfish from that water source.

4-169. Do not eat shellfish not covered by water at high tide.

CRUSTACEANS

4-170. Freshwater shrimp range in size from 1/16 inch up to 1 inch. They can form rather large colonies in mats of floating algae or in mud bottoms of ponds and lakes.

4-171. Crayfish are akin to marine lobsters and crabs. Their hard exoskeleton and five pairs of legs—the front pair having oversized pincers—can distinguish them. Crayfish are active at night, but can be located in the daytime by looking under and around stones in streams. They can also be found by looking in the soft mud near the chimney-like breathing holes of their nests. Crayfish can be caught by tying bits of offal or internal organs to a string. When the crayfish grabs the bait, pull it to shore before it has a chance to release the bait.

4-172. Saltwater lobsters, crabs, and shrimp can be found from the surf's edge out to water 33 feet deep. Shrimp may come to a light at night where they can be scooped up with a net. Lobsters and crabs can be caught with a baited trap or a baited hook. Crabs will come to bait placed at the edge of the surf where they can be trapped or netted. Lobsters and crabs are nocturnal and most easily caught at night.

WARNING: All freshwater crustaceans, mollusks, and fish must be cooked. Fresh water tends to harbor many dangerous organisms, animal and human contaminants, and possibly agricultural and industrial pollutants.

CHAPTER 5

FIRE

This chapter discusses uses and principles of fire. It also describes how to construct a fire through different phases and using different construction methods to meet specific needs. The chapter closes with an in-depth discussion on fire-starting using man-made and primitive methods.

FIRE BASICS

5-1. A properly constructed and utilized fire is one of the best tools an isolated person has. The use of fire supports several different survival needs and provides many other benefits. Fire—

- Provides several ways to signal recovery forces with light and smoke and increases the isolated person's presence in the area with multiple well-placed fires. Fires allow personnel to multi-task while fire acts as a passive signal to recovery forces.
- Provides a sustained source of heat that reduces energy required for the body to maintain its temperature. Consequently, isolated persons require fewer calories and less food.
- Enables the most effective method of water purification, boiling.
- Dries wet clothing and prevents cold injuries such as immersion foot, frostbite, and hypothermia.
- Provides coals and flame to improvise tools.
- Emits heat and smoke to preserve and cook food for consumption.
- Repels mosquitoes and dangerous animals.
- Provides the requisite heat to sterilize items for medical purposes.
- Boosts morale by providing comfort, peace of mind, entertainment, and companionship.

5-2. Isolate persons will require knowledge of how to build, use, and maintain a fire under the most austere and inhospitable conditions. Fire is used with caution to prevent damage and injury to persons and their surroundings. Furthermore, before building a fire you should weigh the benefits of fire against the negative potential effects—such as the chances that the light, smoke, and smell may be detected by the enemy, that it might cause a forest fire, and the danger of burns or carbon monoxide poisoning if a fire is built in a shelter or other enclosed space.

FIRE BURNING PRINCIPLES

5-3. To build a fire, it helps to understand the following basic principles of a fire:

- A typical solid or liquid (non-gaseous) fuel does not burn directly. When heat is applied to a fuel, it produces a gas. This gas, combined with oxygen in the air, burns as long as there is fuel to supply the gas.
- Understanding the fire triangle is very important in constructing and maintaining a fire. The three sides of the triangle represent heat, oxygen, and fuel. If any of these are removed, the fire will go out. The correct ratio of these components is very important for a fire to burn at its greatest capacity (see Figure 5-1). Manipulating the structure of a poorly burning fire is an effective way to adjust the ratio of heat, oxygen, and fuel to reach a desired state.

Figure 5-1. Fire Triangle

SELECTING AND PREPARING A FIRE SITE

5-4. Determining a good site to build a fire is an often overlooked but important step in effectively using a fire. Isolated persons should select a site that provides a good foundation for building a fire and prepare it to support their needs. Considerations include—

- Terrain and climate.

- Materials and tools available for fire building.
- Time available for fire building.
- Expected duration of isolation.
- Purpose(s) for the fire.
- Security risks that a fire may cause.
- Requirement to direct heat in a desired direction.

5-5. Fire sites should meet the following basic criteria:

- Flat, dry area to construct fire and accomplish fire-related activities.
- Protected from the wind.
- Located near isolated person's shelter or living area.
- Near a fuel supply to maintain the fire.
- Away from snow-laden trees.

5-6. To prepare a fire site—

- Clear the ground of fallen leaves, vegetation, and duff[1]; expose 3 feet (1 meter) in diameter of soil at a minimum to prevent the fire from spreading along the ground.
- Porous and wet rocks are removed from the fire site, as they are at risk of exploding from the heat of the fire.
- Level the area that the fire is to be constructed upon, if sloped, to prevent fuel from sliding off the fire.
- In deep snow where the soil cannot be reached, green logs are used as a base to build the fire on to prevent the coals from melting a hole through the snow pack and away from you. In the cold, green logs with a diameter up to the size of a soda can are easily broken. Pieces are stacked in two layers, the top being perpendicular to the bottom to a size that contains the entirety of the intended fire.
- Construct a fire reflector if desired; this task is more easily accomplished during the preparation phase before the fire becomes hot.

FIRE REFLECTOR

5-7. Fire reflectors are used to maximize the heat produced from the fire by directing it in a desired direction and protecting the heat from being displaced by the wind. Reflectors can be constructed of nonporous rock or a stack of logs (most effectively made of green wood—see Figure 5-2 on page 171). When located at the opening of a shelter, the heat is directed into the enclosure to elevate the temperature of the living area.

---

1. Duff: the layer of organic, decomposing material lying immediately above the soil layer.

Figure 5-2. Types of Firewalls

Straight firewall

L-shaped firewall

BUILDING A FIRE

5-8. Once the fire site is selected and prepared, the fire-building process can begin. The first step is to procure a base and a brace (see Figure 5-3 on page 172) which are used to initially establish and control the fire. The base is a flat piece of material, usually wood or a nonporous rock (porous rocks contain water and, when heated, can explode). The base is used to keep moisture-susceptible tinder and kindling off the wet ground and provides a clear area to ignite tinder. The brace is employed across the base to help control oxygen into either the tinder, kindling, or fuel being worked. It is about the size of a person's forearm. A broken branch can typically be found and employed as a brace.

Figure 5-3. Base and Brace

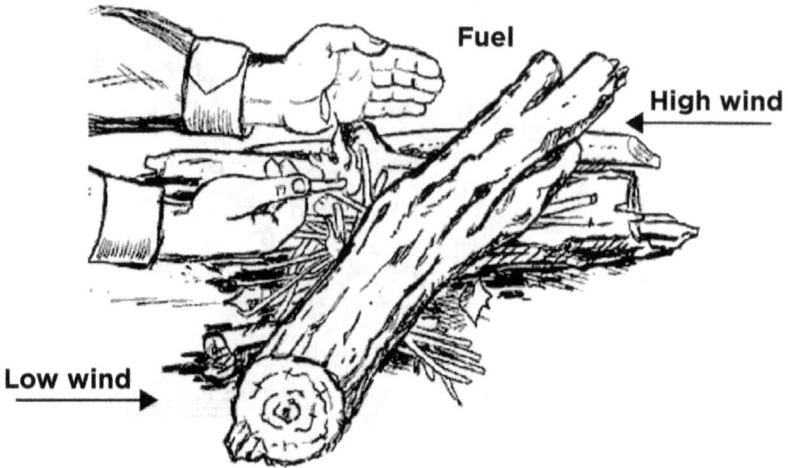

5-9. After preparing the base and brace, fuel is procured and prepared to start and maintain the fire. Fire building occurs across 3 phases. If one phase is skipped even in ideal circumstances, it is likely that the fire will fail due to not having the appropriate mix of oxygen, heat, and fuel. The 3 phases are tinder, kindling, and fuel, each representing an increase in fuel size appropriate for heat and oxygen available to support it. Figure 5-4 provides examples of tinder, kindling, and fuel.

Figure 5-4. Tinder and Kindling Examples

TINDER

5-10. Tinder is used during the first phase of fire building. A dry material can be ignited by a small spark, coal, or flame. Tinder is normally comprised of fine fibers or is petroleum-based and is either natural or man-made. Isolated persons can procure natural tinder, when damp, and store it between layers of clothing to use heat radiated by the body to dry it prior to use. When using fibrous tinder to start a single fire, you should collect enough to fill your patrol cap or your hands with fingers cupped at a minimum. Natural tinder examples include—

- Birch bark—shredded inner bark known as heartwood from cedar, chestnut, red elm trees.
- Fine wood scrapings or shavings (heartwood preferred).
- Dead dry grass, ferns, moss, lichen, and fungi.
- Dry powdered sap from the pine tree family (also known as pitch).
- Fine pitch wood scrapings or shavings.
- Crushed fibers from a dead plant.
- Pine pitch globules mixed with fibrous material.
- Seed down (milkweed, cattail, thistle).
- Feather sticks (wood shavings cut into and still connected to a piece of wood).

5-11. Man-made tinder examples include—

- Cotton balls (most effective when coated with a petroleum-based paste).
- Lint.
- Paper.
- Foam rubber.
- Hexamine.
- Tryoxine.
- Commercial petroleum-based pastes.
- Steel wool.
- Gunpowder.
- Candles.
- Solid-state fuel tabs.

5-12. A sufficient amount of tinder must be used to generate a hot enough flame to ignite kindling in the next phase of fire building. Smaller quantities of petroleum-based products are used as they burn much longer with a central burn point. To ignite some petroleum-based tinder and tree sap, it may be necessary to combine it with some fibrous tinder to catch the flame for ignition.

5-13. Tinder is laid on a dry base to avoid the fine fibers from soaking up moisture and rendering it unusable. When laid on the base it rests up against the brace in a mound. The tinder is lighted from its bottom edge away from the brace. The initial flame is small and is most easily able to spread in an upward sloping direction creating a more reliable, quicker, and hotter flame to engulf the tinder and ignite kindling.

KINDLING

5-14. Kindling is the second phase of fire building. Kindling is added once the tinder is burning strong and mostly engulfed by flame. It is laid by the handful on, and perpendicular to, the top of the brace and over the burning tinder. It should be laid over the tinder fire so that some spaces are left in the kindling for the tinder fire to burn through: if the kindling was laid over the tinder like a blanket, the fire would likely not receive enough oxygen to continue burning, would not be able to ignite the kindling, and would die out. The second handful of kindling is laid perpendicular or diagonal to the first laid kindling to maintain small pockets for the fire to burn through and grow.

5-15. Kindling has a high combustible point similar to tinder. The kindling phase is used to transition to large long-burning fuel by increasing the size of the fire, and creating an initial coal base that increases heat output and the fire's longevity.

5-16. Examples of kindling include—

- Dead, dry, small twigs and plant stalks; you should consider that during the winter months live deciduous vegetation that is dormant may appear dead. A quick check of the cambium layer—just under the bark—will indicate whether it is alive or dead.
- Dead, dry, thinly shaved pieces of wood, bamboo, or cane (always split bamboo as un-split sections can explode).
- Dead coniferous needles.
- Small branches low to the ground on the underside of coniferous trees that are dead from lack of sun (a single branch often has all three sizes of kindling on it that simply need to be broken down for use).
- Some plastics such as the spoon from an in-flight ration.
- Wood which has been soaked or doused with flammable materials such as wax, insect repellent, petroleum fuels, or oil.
- Strips of petrolatum-impregnated gauze from a first aid kit.
- Pitch wood sticks.
- Larger pieces of wood split into kindling-sized pieces.

5-17. The inner wood of dead branches and small diameter trees is the driest because it is not exposed to the weather. Wood that is split to expose the inner wood grains burns faster and hotter than wood pieces that are covered by bark, which naturally protects trees from fire and collects moisture. Wood less than 3 inches in diameter and approximately 1 to 2 feet in length can easily be split using a strong fixed-blade survival knife and a sturdy forearm sized branch. The branch is used to drive the knife down the length of the wood along the grain (batoning). Wood with the fewest knots is selected for splitting. The split is repeated until the desired size is reached.

5-18. The following method is used to split wood with a fixed-blade survival knife:

Step 1 Set the piece to be split upright on the ground, or on a sturdy base, and hold it in position.

Step 2 In the non-primary hand, hold the knife, blade down, on the top of the wood piece, the tip should extend well past the edge of wood and slightly into the air so that the knife is at about a 60-degree angle to the wood piece.

Step 3 Place pressure on the handle of the knife to ensure the blow of the branch onto the tip end of the knife drives the blade into the wood (if proper pressure is not applied the energy from the blow will instead result in popping the handle into the air)

Step 4 Strike a blow with the branch against the tip end of the knife.

Step 5 Repeat until the wood piece is split through; continue splitting each piece into smaller pieces to reach the desired size.

Note: Ensure that your knife is sufficiently sturdy before using this method to split wood. Batoning can easily cause the blade of a knife without a full tang (a full-tang knife is one where the metal of the blade extends down the full length of the knife's handle, conferring greater strength) to separate from the handle or, for a folding knife, cause the blade locking mechanism to fail. Some perfectly good knives designed for combat or other purposes rather than survival or camp use will fail if used for batoning. If in any doubt about your knife's strength, use a different implement rather than risk losing the use of your knife.

5-19. Preparing kindling in several different diameters that range from pencil-lead, to pencil, and then to thumb size will make fire-building easy when laid over the tinder in that order. Kindling that is approximately 12~18 inches in length is the easiest to apply and manipulate when building a fire. Collect at an amount the diameter of a soda can of each size (at the very minimum), to ensure enough kindling is available to increase the fire's heat to ignite the fuel in the next phase.

FUEL

5-20. Wood used as fuel should be the diameter of the forearm at a minimum. Wood is the most abundant source of fuel in most environments; however, there are many other types of fuel an isolated person can use. Unlike tinder and kindling, fire fuel does not have to be kept completely dry if the fire is established well enough to dry the damp fuel before burning. However, the more moisture fuel contains the more white smoke the fire will produced as it burns it off. The type of fuel will determine the amount of heat and light the fire produces. Dry, split hardwood trees (such as oak, hickory, and ash) are less likely to produce excessive smoke and provide more heat than softwoods (pine, spruce, and fir) due to their higher density. This density however makes them more difficult to break and split into usable pieces than softwood. Softwood burns faster and produces more smoke unless a large flame is maintained. They are often straight grained, have fewer knots, and are easy to split.

5-21. Effective fuel sources include dry, standing, dead wood and dry, dead branches (those that snap when broken). Standing or leaning dead wood is usually dry even in rainy climates as its vertical position sheds most rain. In contrast, rotten wood, most of which is found lying directly on the ground, is of little use since it smolders and is very difficult to ignite. It produces a large amount of smoke, which can be useful in a signaling situation, but is a poor fuel.

CAUTION: Highly flammable liquids should not be poured on an existing fire. Even a smoldering fire can cause liquids to explode and cause serious burns.

5-22. Dead wood is easy to split and break. It can be pounded on a rock or wedged between trees 2~3 feet apart, bending it until it breaks. A large survival knife can also be used to split the wood to reach the heartwood, which is typically the last to rot. As a last resort green wood, which is found almost anywhere, can be made to burn if finely split and mixed evenly with dry, dead wood. In extremely cold weather, moisture is pulled from green wood, which then becomes a viable fuel. Wood used as fuel must be free of poisonous plants such as poison ivy or poison oak. These plants can cause respiratory distress if burned and inhaled. In areas void of wood, other fuels can be used:

- Dry grasses can be twisted into bunches.
- Dry bamboo can be used as a fuel after it is split to open its sections (which, if not split, could explode); you must beware of large splinters.
- Dead cactus and other plants are available in deserts.
- Dry peat moss can be found along the surface of undercut stream banks and in dry swampy areas.
- Dried animal dung, animal fats, and sometimes even coal can be found and used as fuel.
- Blubber from marine animals can be used as a fuel source.

- Oil, diesel, gasoline, and other high-performance fuels can be used when mixed with sand to slow their burn rate and prolong their burn time. This makes them a safe fuel alternative.
- Vehicle tires burn well but are difficult to ignite and produce chemicals that should not be inhaled.

CAUTION: Petroleum-based fuels will produce thick, black smoke when burned. Inhaled petroleum products can be harmful to your health and respiratory system. Do not prepare foods directly over petroleum-based fires due to the toxins produced. Ensure proper ventilation when utilizing these types of fuels.

FIRE LAYS

5-23. Lay the fuel in a manner consistent with your purpose as listed in Paragraph 5-1. A fire lay is a structure constructed by the manner in which fuel, typically wood, is placed to focus heat or flame, accommodate personnel, provide light, or use fuel. Selecting the correct fire lay to support specific tasks will make the fire a more effective and efficient tool. Commonly used fire lays include—

TEPEE

5-24. Tepee fires are made by arranging the tinder and a few sticks of kindling in the shape of a tepee or cone. The center can be lighted after the structure is built or the structure can be built around a small fire. As the tepee burns, the outside logs will fall inward, feeding the fire. This type of fire burns well even with wet wood. (See Figure 5-5 on page 179 for an example).

LEAN-TO

5-25. Lean-to fires are constructed by pushing a green stick into the ground at a 30-degree angle. The end of the stick is pointed in the direction of the wind. Tinder is placed deep under the lean-to stick. Kindling is then placed against the lean-to stick. The tinder is then lighted. As the kindling catches fire from the tinder, more kindling and later fuel is added. (See Figure 5-5 on page 179 for an example).

CROSS-DITCH

5-26. Cross-ditch fires are constructed by digging an "X" about 12 inches in diameter, and 3 inches deep into the ground. Tinder is placed in the middle of the cross. A kindling pyramid is built above the tinder. The shallow ditch allows air to sweep under the tinder to provide a draft. (See Figure 5-5 on page 179 for an example).

PYRAMID

5-27. A pyramid fire is built by placing two small logs or branches parallel on the ground. A solid layer of small logs are then placed across the parallel logs to make a platform. Three or four more layers of logs are added, each layer smaller than the last and perpendicular to the layer below it. A starter fire is then built on top of the pyramid. As the starter fire burns, it will ignite the logs below it. This creates a fire that burns downward, requiring no attention during the night (see Figure 5-5 on page 179 for an example).

LOG CABIN

5-28. As the name implies, this lay looks similar to a log cabin. Log cabin fires are built by placing layers upon layers of perpendicular fuel wood. This fire creates a great amount of light and heat, primarily because of the amount of oxygen which enters the fire. The log cabin fire creates a quick and large bed of coals and can be used for cooking or as the basis for a signal fire.

LONG FIRE

5-29. The long fire can be made as a trench, the length of which is laid in a direction that will take advantage of existing wind. It can also be built above ground by dragging two logs (green is better) parallel to each other to hold the coals together. These logs should be at least 6 inches in diameter and situated so cooking utensils can rest upon the logs. Two 1-inch-thick sticks can be placed under both logs, one at each end of the long fire. This is done to allow the coals to receive more air. Long fires allow larger groups of personnel near the heat and enable coals to be moved to one end for cooking, with a fire at the other.

STAR FIRE

5-30. Star fires are built by placing wood pieces or logs in the shape of spokes in a wheel so that the inner ends burn in the fire. As the ends burn, the remaining fuel from each log is pulled closer to the fire. The star fire is used when conservation of fuel is necessary or a small fire is desired. It also allows you to use larger pieces of fuel that you can otherwise not break down into smaller pieces suitable for other fire lays. This fire must be constantly tended. Hardwood fuels work best for star fires.

Figure 5-5. Fire Lays

Teepee

Lean-to

Cross-ditch

Pyramid

5-31. When you are done with a fire it should be extinguished using materials such as water, entrenching tools, dirt or sand, or sturdy sticks that are prepared in advance. A fire is extinguished by removing one of the fire triangle elements. Fires with non-petroleum-based fuel can be extinguished with an adequate amount of water. Large pieces of burning wood can be separated and the coals can be raked off with a tree branch. Some burning fuel can be broken apart and scattered over bare soils. Dirt or sand works well to extinguish petroleum-based fires. Isolated persons may even be able to retrieve a chemical-based fire extinguisher from their aircraft or vehicle for this purpose.

CAUTION: When leaving an area after having a fire, ensure that the ground under where the fire was built is cool to the touch.

FIRE STARTING TECHNIQUES

5-32. Fires are started from the upwind side to protect the fragile fire-starting heat source. Several different techniques can be used to ignite tinder. These techniques support either modern or primitive methods. Isolated persons determine the best technique to use after analyzing issued and natural resources available to start the fire. While primitive methods allow for greater flexibility (since they can be improvised from the natural environment), they generally require more skill and time to execute. Modern methods are the quickest and easiest method to start a fire, but they can only start a certain number of fires until the resource is exhausted. However, some modern methods, such as a metal match, can be used hundreds to thousands of times over depending upon the skill of the user.

MODERN METHODS OF FIRE BUILDING

5-33. Modern fire-starters provide an advantage because they are specifically constructed for igniting fire, are durable, and their performance can be easily repeated. Other man-made devices can be used as an improvised fire-starter with minimal effort. Modern methods are often easier for injured isolated persons to utilize when their movement is limited. Modern fire-starters include are discussed in more detail in Paragraphs 5-34~5-56.

Matches

5-34. Ensure that your matches are waterproof, stored in a waterproof container, and have a striker pad. Take intentional precautionary measures to protect lit matches from wind—each match that fizzles out or is blown out by wind represents one fewer fire that you can start in the future. Matches can quickly become damp due to their size and exposed wood grain. When damp, their effectiveness is reduced causing them to burn more weakly or not at all. Matches that have wax coated tips can be purchased commercially. You can also dip the match heads in melted wax to protect them from moisture. Strike-anywhere matches are the most adaptable and desirable, as the use of safety matches (which require a special surface for striking) demands that you also keep the striking surface entirely free of moisture.

5-35. To light a match the head is raked, with evenly applied pressure through the strike, over a rough surface compatible with the match type. Once the match head is ignited, it is held with the match head pointing down at an angle so the flame can begin to climb the matchstick and gain strength. It is critical that the flame is protected from the wind during this step. Once the initial flame gains strength, the match is held near the bottom edge of tinder to ignite it. Caution is used to avoid snuffing out the small match flame by placing it into or tightly against the tinder and smothering it.

Lighters

5-36. Lighters are used to ignite tinder similarly to matches. Lighters come in butane and liquid-fuel types. These can be used for a direct and sustained flame to light the tinder. Butane lighters provide better wind resistance, and are a much hotter and more concentrated heat source than liquid fuel types. For these reasons, they are a better tool for starting a fire. However, liquid fuel types can last longer and create more fires if tinder is prepared properly. Under extreme cold temperatures, lighter fuel can freeze up and not function. In such situations, body heat can be used to warm the lighter. If the fuel runs out of a liquid-lighter, the sparking mechanism can be used similarly to a metal match.

Metal Match

5-37. The metal match is the most useful and versatile issued survival fire-starter when personnel are trained in its effective use. However, quality varies between metal matches and affects how easily they spark, and how hot and long the sparks burn. Personnel should conduct operational tests of metal matches they purchase or are issued to ensure they are effective. Metal matches are reliable and not effected by wet or dry conditionals, cold or hot weather, evaporation or age, and do not malfunction. Metal matches should be protected from corrosion like all metal tools. They are limited only in that they require a piece of metal with a good edge, 90-degree is preferable, to produce hot sparks. Most commercial metal matches include a striker to do this, but the back of a fixed-blade knife, tang end of a folding knife blade, or other similar tool can also accomplish the task just as well.

5-38. Depending upon its size, the quality of the metal match, and the skill of the operator, a single match can start thousands of fires. Metal matches are made of the same type material used for flints in commercial cigarette lighters. Some also have magnesium strips attached that can be scraped into tinder to create a higher flash point to more easily ignite the tinder.

5-39. Holding the striker at a 90-degree angle against the stick typically produces the best sparks. Good sparks from a metal match burn hot, fast, and can compensate for some moisture in tinder. The following are instructions for using a metal match to start a fire:

- Hold the stick in the non-primary hand and striker in the primary hand.
- Grasping the stick at the top, place the bottom end of the stick firmly on or near the edge of tinder setting on a secure base.
- Hold the edge of the striker at a 90-degree angle to the top of the stick.
- Drive the striker downward with evenly applied pressure through the strike to near the bottom of the stick. This will throw a substantial amount of sparks into the tinder of done correctly.
- Terminate the downward strike prior to the striker disturbing the tinder or recently thrown sparks to avoid snuffing them out. To avoid this, the far end of the striker (for example, the tip of a knife) can be slightly angled downward during the strike so that the tip stops the strike when it is driven into the base, rather than the

entire flat side of a knife stamping out the sparks and disturbing the tinder.

Batteries

5-40. Batteries found in vehicles, aircraft, or electronic equipment can be used as a field-expedient fire-starter. Two insulated wires are required in addition to the battery to utilize this method. Once wire is connected to the positive post of the battery and the other wire is connected to the negative post. The two ends are touched to a third piece of uninsulated wire. (See Figure 5-6). A short in the wire will result and cause a spark or a superheated piece of metal to ignite the tinder. Similarly, fine-grade steel wool can be stretched between the two posts of the battery causing the steel wool to heat up and produce a coal.

Figure 5-6. Starting a Fire with a Battery

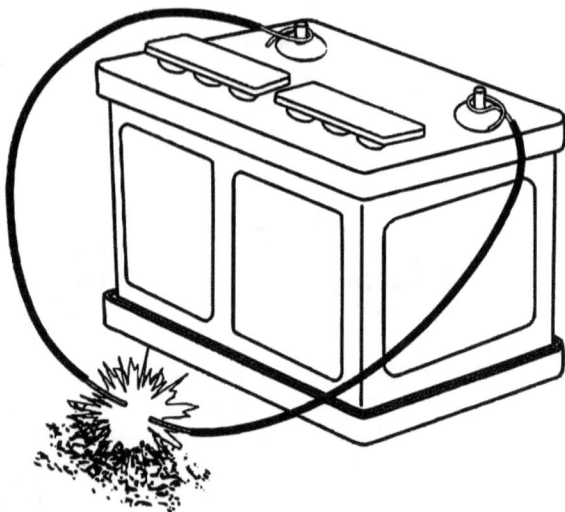

CAUTION: Be careful to keep any open flame away from the battery. Certain batteries produce explosive hydrogen gases that could result in injury.

Convex Lens

5-41. Use this method only on bright, sunny days. The lens can come from binoculars, a camera, telescopic sights, or magnifying glasses. Angle the lens to concentrate the sun's rays on the tinder (see Figure 5-7 on page 183). Hold the lens over the same spot until the tinder begins to smolder. Gently blow or fan the tinder into a flame.

Figure 5-7. Using a Convex Lens as a Fire-starter

Flashlight Reflector

5-42. A flashlight reflector can be used to start a fire on a bright, sunny day by placing the tinder in the center of the reflector where the bulbs are usually located. The tinder is pushed up from the back of the hole until the hotter light is concentrated on the end and smoke results. If a cigarette is available it can be lit and the coal applied to tinder.

PRIMITIVE METHODS OF FIRE BUILDING

5-43. Primitive fire-starters are developed using the resources of the natural environment. They can be time-consuming to construct and require greater energy, patience, and skill to learn in comparison to modern fire- starters. However, isolated persons that are unprepared or separated from their equipment can effectively use these methods to support survival tasks. Personnel that practice these skills before isolation are much more likely to be successful in constructing and using primitive fire-starters.

Flint and Steel

5-44. The type of rock most commonly used in fire starting is flint or any type of rock in the flint family such as quartz, chert, obsidian, agate, or jasper. Other stones may also work. The main criteria are for a rock to be high in silica content and harder than steel. The direct-spark method is the easiest primitive method to use. Strike a flint or other hard, sharp-edged rock with a piece of carbon steel (stainless steel does not produce a good spark). See Figure 5-8. This method requires a loose-jointed wrist and practice. When the tinder catches a spark, blow on it. The spark will spread and burst into flames. Charred cloth is the preferred tinder for this method. The process for making charred cloth involves small pieces of cotton material, a small sealable metal container. Charred cloth can be made using the following instructions:

- Place a nail-sized hole in the container lid.
- Put cotton material loosely in layers inside the container and put the lid on.
- Set the container near the edge of the fire on top of hot embers.
- Once heated, smoke will escape the lid; an indication that the process is working.
- Once the smoke stops, the container is removed from the fire and allowed to cool.
- After it has cooled, the container is opened away from any exposed flame and stored in a water-resistant container.

Figure 5-8. Flint and Steel Fire-starter

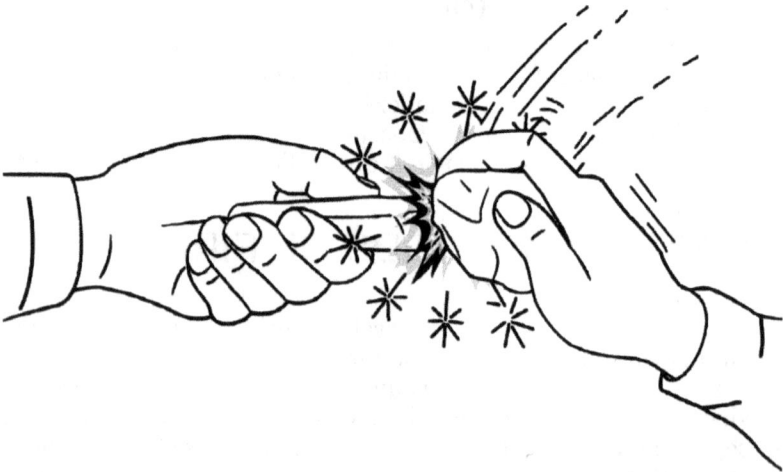

Fire Plow

5-45. The fire plow is a friction method of ignition (see Figure 5-9 on page 185). To use this method, cut a straight groove in a softwood base and plow the blunt tip of a hardwood shaft up and down the groove. The plowing action of the shaft pushes out small particles of wood fibers. As more pressure is applied on each stroke, the friction ignites the wood particles.

Figure 5-9. Fire Plow

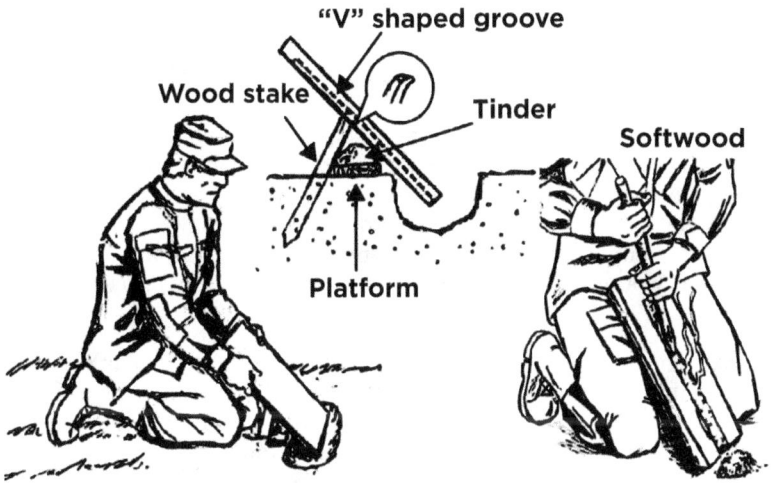

"V" shaped groove

Wood stake

Tinder

Softwood

Platform

Bow and Drill

5-46. The bow and drill method is simple in concept but requires significant preparation initially. You will need to improvise several parts including a socket, drill, fireboard, fire pan and bow from materials located in your nearby surroundings. Depending upon available resources, isolated persons may be able to use man-made elements to supplement natural materials in constructing a bow and drill.

Socket

5-47. A socket is an easily grasped stone or piece of hardwood with a slight depression in one side. The depression may need to be created. The depression in the socket is used to hold the top of the drill in place as it is rotated in the V-shaped cut of the fireboard. Soap or grease can be used in the socket to reduce friction. Rock is the best material from which to build a socket. Alternatively, pitch wood and very hard wood types will last the longest as sockets before the top of the spindle wears through it. Soft woods can be used, but will need to be replaced more often.

Drill

5-48. The drill should be a straight, seasoned hardwood stick about ¾ inch in diameter and 10~12 inches in length. The top end is rounded so it fits into the depression of the socket. The lower end is blunt so it will produce greater friction. Octagonal sides create greater friction than smooth sides when spinning. Drills that have a slight hourglass shape or groove in the middle will assist in keeping the bow in the middle of the drill and facilitate strokes that are more efficient.

Avoid any resinous woods as they will glaze over and hinder the process.

Fire Board

5-49. The fireboard should be a seasoned soft wood board. Examples of woods that work well as fireboards are yucca, tamarack, aspen, balsam fir, basswood, poplar, cypress, cottonwood, alder, red cedar, and willow. The fireboard should be about 1 inch thick and 4 inches wide. Cut a depression about ½~¾ inches from the edge on one side of the board. On the underside of the fireboard, make a V-shaped cut from the edge of the board to the depression. Again, avoid any resinous woods as they will glaze over and hinder the process.

Fire Pan

5-50. The fire pan can be thin bark, leather, or a flat plank of wood placed under the fireboard to catch the wood dust and coals produced from the drill. Once a coal has formed, the fire pan can be picked up with one hand to transfer the coal to the tinder bundle in the other hand. The fire pan also acts as a base to protect the fireboard against ground moisture.

Bow

5-51. The bow should be a strong, green stick about 3/4 inch in diameter with a length of about three feet, that should extend long enough to go from your finger tips to your shoulder. The stick should be flexible enough to bend into a bow, appearing similar to an archery bow. The type of wood is not important, but must be strong. The bowstring can be made from some type of cordage or leather. The cordage is tied from one end of the bow to the other, without any slack. When in use, the bowstring is twisted around the drill once to rotate it as it spins or drills into the fireboard producing the friction needed to create a coal.

Tinder

5-52. Tinder used for a bow and drill fire must be very dry, fine fibrous material made into the shape of a birds nest. Grass, lichen, cotton, cattail duff, or cedar bark torn into small fibers is preferred. A large amount of completely dry tinder should be collected and prepared on a dry base near the fire-site (and the bow and drill operation) in order to be able to quickly transition a coal from the fireboard to the tinder when ready.

Operating a Bow and Drill

5-53. The base, brace, tinder, and kindling are prepared prior to operating a bow and drill. The tinder is placed near the fireboard on a base to keep it from ground moisture. The following steps are taken to operate a bow and drill fire-starter (see Figure 5-10 on page 188):

Note: These instructions assume you are right-handed—for left-handers, swap the sides.

- Place the left foot on the fireboard to secure it in place and the right knee on the ground.
- One turn of the bowstring is made over the center of the drill.
- The bottom of the drill is placed in the fireboard notch and the tip is placed in the socket held by the left hand.
- Apply pressure on the drill from the socket and make long, smooth strokes of the bow at a steady moderate speed; each stroke makes full use of the available string as longer strokes produce better friction.
- Once a coal is produced in the mound of black wood dust on the fire pan, it begins to lightly smoke, at which time the fireboard is removed and the bird nest shaped tinder is place over the coal.
- The fire pan and bird nest are then both flipped so the bird nest is on the bottom and fire pan is on top, and the fire pan is then removed.
- The bird nest is then lightly closed around the coal and lightly waved through the air or lightly blown on with long steady breaths to heat the coal and surrounding tinder.
- The amount of smoke coming from the inside of the bird's nest will continue to increase until it bursts into flame, at which time it is placed on the base where kindling is added.

Figure 5-10. Bow and Drill Fire-starter

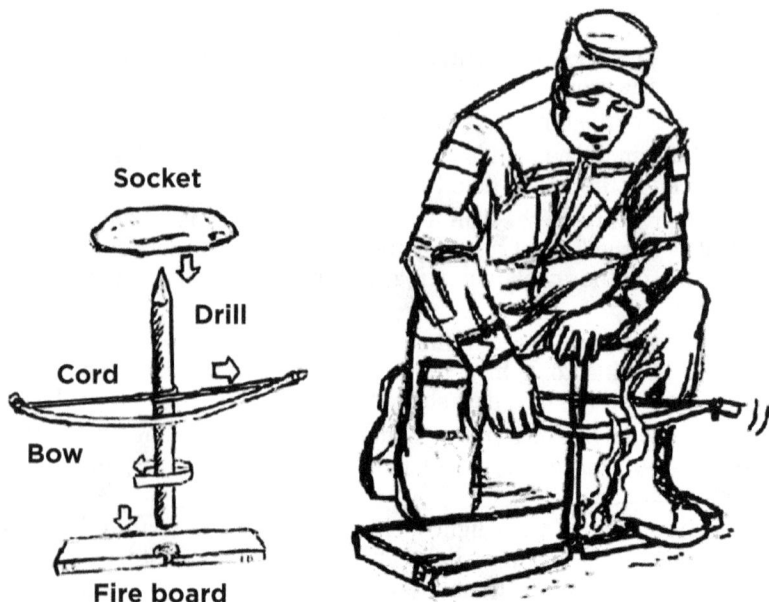

Hand Drill

5-54. A hand drill is a simpler method than the bow and drill in that it requires less preparation. However, its operation is more difficult for some. The lower rate of rotation between the hands makes it easy to use in hot and dry climates where the wood is very dry. The drill can be made of dried yucca, elderberry, and other similar wood. The fireboard is constructed of the same materials and dimensions as the one used for bow and drill fires. To operate a hand drill, pressure is applied to the top of the drill while rotating the hands to spin it in short bursts equal to the length of the hands. The hands will gradually slip down the length of the spindle. (See Figure 5-11 on page 189). When near the bottom of the drill, the hands are repositioned and the action is repeated.

Figure 5-11. Hand Drill Fire-starter

FIRE BUNDLE

5-55. Isolated persons may need to move to meet their needs. When they do, transporting coals from a previous fire with them may be the easiest method to start a new fire. Doing so will also conserve potentially limited fire-starting resources. This can be accomplished by constructing a fire bundle to carry smoldering embers. A tin can, animal horn, hardwood box, and birch bark pouch can be used for this purpose.

5-56. Embers carried in a well-built fire bundle can last for several days. For example, a fire bundle built from birch bark can be constructed by laying out a large piece of bark flat, adding a layer of wet material, then rotten wood. The coals are placed in the rotten wood and the bundle is rolled shut. (See Figure 5-12 on page 190). The bundle should be checked occasionally to maintain the coals and ensure enough oxygen reaches them to prevent them from being extinguished.

Figure 5-12. Fire Bundle

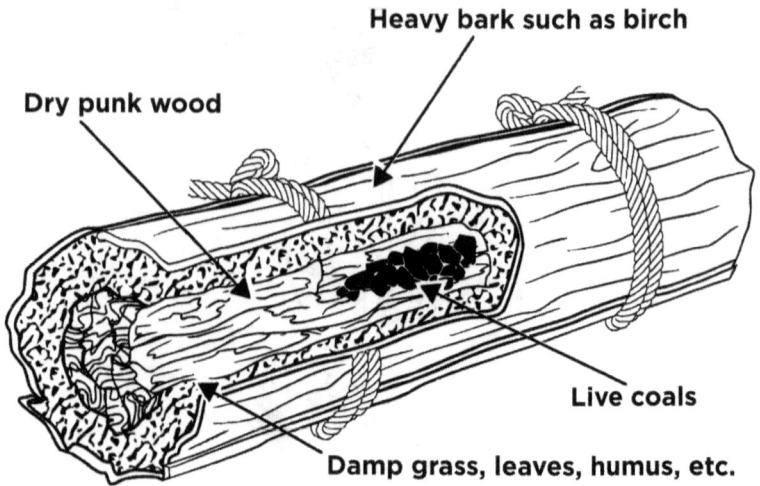

Heavy bark such as birch

Dry punk wood

Live coals

Damp grass, leaves, humus, etc.

CHAPTER 6

SHELTER AND CLOTHING

A shelter can protect you from the sun, insects, wind, rain, snow, hot or cold temperatures, and enemy observation. It can give you a feeling of wellbeing and help you maintain your will to survive. In some areas, your need for shelter may take precedence over your need for food and possibly even your need for water. For example, prolonged exposure to cold can cause excessive fatigue and weakness. This chapter discusses clothing and both man-made and natural shelters.

PRIMARY SHELTER

6-1. Your primary shelter in a survival situation will be your uniform. This point is true regardless of whether you are in a hot, cold, tropical, desert, or arctic situation. For your uniform to protect you, it must be in as good a condition as possible and be worn properly.

6-2. Clothing needs to keep the isolated person dry, warm or cool, out of the sun, and away from bugs. The foundation of your uniform is your footwear. Personnel should ensure they have broken in their boots before wearing them on a mission to ensure they do not get blisters. Good socks also provide protection from blisters. Socks available to personnel are typically made of polypropylene, cotton, and wool.

6-3. The key in choosing the right sock is to ensure the material meets your needs. For example, wool socks have great insulation value; wool is animal hair of hollow fibers and it retains heat. It absorbs moisture but retains it in its fibers, and can absorb about fifty percent of its weight in water before it feels cold and wet. Wool is hard to dry out but retains its insulating properties when wet. Wool is flame resistant and is more forgiving than other fabrics when drying by fire. The wool fiber is barbed and spear-shaped and can cause an itchy feeling. However, new manufacturing processes and different breeds of wool help with this problem.

6-4. Cotton socks keep the foot cool and wick moisture away from the foot; once wet, cotton will lose about ninety percent of its insulation value and is very difficult to dry out. In dry, warm climates, cotton socks should be the isolated person's sock of choice because they can be extremely abrasion- and ultraviolet-ray resistant, and they can help keep feet cool. Polypropylene socks resist absorbing moisture. The material transfers moisture across the material from the skin onto other material or the air where it can dry out.

6-5. Polypropylene makes an excellent base layer. One down side of polypropylene is that it is so effective at wicking away moisture it can cause the body to overwork itself and use more energy than other fabrics. This fabric also retains body odor and has a low melting point, so caution should be used when drying it around a fire. Personnel should carry extra socks and rotate them as needed to keep their feet dry. Foot powder can assist in this process.

6-6. Clothing fibers can be degraded by sand and soil because of the abrasive nature of their grains. Additionally, isolated personnel may develop blisters or a hot spot causing injury when sand- or dirt-covered clothing rubs against their skin. They must make every effort to attain a high standard of cleanliness including the use of platforms to avoid putting clothing and equipment directly on the ground (but not at the expense of security).

6-7. If clothing gets dirty, clean it and keep it dry. The "COLDER" (**C**lean, **O**verheating, **L**oose **L**ayers, **D**ry, **E**xamine, and **R**epair) principle provides a foundation for care and wear of the uniform while isolated and is further described in the following:

- **Clean.** Keep clothing clean. This principle is always important for sanitation and comfort. In winter, it is also important from the standpoint of warmth. Clothes matted with dirt and grease lose much of their insulation value. Heat can escape more easily from the body through the clothing's crushed or filled-up air pockets. Dirt within the fibers will cut or tear the clothes, causing them to wear out prematurely. In a survival situation, it may be impractical to wash clothing; therefore, take the necessary precautions to prevent clothing from becoming soiled.

- **Overheating.** Avoid overheating. When personnel get too hot, they sweat and their clothing absorbs moisture. This affects their warmth in the following ways: dampness decreases the insulation quality of clothing and as sweat evaporates, the body cools. Adjust your clothing so that you do not sweat. Do this by partially opening your parka or jacket, by removing an inner layer of clothing, by removing heavy outer mittens, or by throwing back your parka hood or changing to lighter headgear. The head and hands act as efficient heat dissipaters when overheated.

- **L**oose **L**ayers. Wear clothing loose and in layers. Tight clothing and footwear restrict blood circulation and invite cold injuries. It also decreases the volume of air trapped between the layers, reducing their insulation value. Several layers of lightweight clothing are better than a single, equally thick layer of clothing, because the layers have dead airspaces between them. This dead airspace provides extra insulation. In addition, layers of clothing allow personnel to take off or add layers to prevent excessive sweating or to increase warmth.

- **D**ry. Keep clothing dry. This is important since wet clothing can conduct heat away from the body up to 25 times faster than dry clothing, depending on the type of fabric.

  - Wear water-repellent clothing as an outer layer if available. It will shed most of the water from rain and snow. Despite the precautions taken, there will be times when isolated persons cannot avoid getting wet. At such times, drying your clothing may become a major problem.

  - As isolated persons move about, they can place wet clothing on their rucksack, exposing the wet clothing to the sun and wind; this will dry the clothing. They can also put damp clothing near their body's core and body heat will began to dry the clothing. If at a stationary location, construct drying racks or a clothesline to hang clothes up to dry.

  - Placing clothing near a fire will dry it quickly. You should make sure to use the three-second rule so you do not put clothing too close to the fire and damage it. The three-second rule involves being able to hold a hand in front of the fire at a distance that, if the back of the hand is exposed to the flame, it is still comfortable to keep it there for three or more seconds. Also, monitor the clothing to ensure that it does not fall into the flame or that the fire does not flame up and destroy the items.

  - Dry leather items slowly. If no other means are available for drying boots, isolated persons can put the boots between their sleeping bag liner and shell. Their body heat will dry the leather. Drying leather with direct heat (such as from a fire) is to be avoided as it will shrink the leather and crack it.

- **E**xamine. Periodically examine clothing for worn areas, tears, and cleanliness.

- **R**epair. Repair any damaged clothing before tears and holes become too large to patch or sew. Patches can be made from cloth, tape, and other materials. Use a sewing kit, improvising one if needed. Sewing kits can be made from such items as bones, plant fibers, parachute cord, and large thorns.

SHELTER CONSIDERATIONS

6-8. Shelter is anything that provides protection from environmental hazards. The environment influences shelter site selection and other factors, which isolated personnel must consider before constructing an adequate shelter. Depending on the environmental conditions, shelter may be the isolated person's number one priority. You can take advantage of naturally occurring features or make a shelter from man-made and natural materials brought along or found in the environment. Consider the type of shelter, location, and time of day when looking to build a shelter. Also consider the location of shelter in relation to temperature, precipitation, and wind. You should consider the time and effort needed to build adequate shelter and protection from the elements. Some environments require shelter to be built off the ground, others will require more insulation, and still others will require the effective use of winds for cooling.

SHELTER SITE SELECTION

6-9. When isolated persons have prioritized shelter as a necessity, they must start looking for it as soon as possible. A good shelter site will contain enough materials to improvise the type of shelter needed as well as be large enough and level enough for you to lie down and protect your equipment. When selecting a shelter site, consider whether the site—

- Provides adequate materials to build the right type of shelter.
- Provides concealment from enemy observation.
- Provides elemental sun, wind, rain, and snow protection.
- Has camouflaged escape routes.
- Is free of dead trees that may fall.
- Is located in an area free from threat of avalanches, rockslides, mudslides, flash-flood areas and high-water marks.
- Is free from insects, reptiles, and poisonous plants.
- May act as a "cold sump" (since cold air sinks, low-lying areas may be cooler).
- Is suitable for signaling and recovery.
- Contains adequate food and water.

6-10. When looking for a shelter site, keep in mind the type of shelter needed. Ask yourself the following questions:

- How much time and effort will I need to build the shelter?
- Will the shelter adequately protect me from the elements?
- Do I have the tools and needed materials to complete the shelter or do I need to improvise?
- How long will I be at this site?

SITE PREPARATION

6-11. Site preparation includes brushing away rocks and twigs from the sleeping area and cutting back overhanging vegetation. Look for loose rocks, dead limbs, coconuts, or other material that could fall on the shelter. Remember thick, brushy, low ground also harbors more insects. Inspect the site for venomous snakes, ticks, mites, scorpions, and stinging ants.

CONSTRUCTION METHODS

6-12. General considerations for the construction of a shelter include the following—

- **Size**. The shelter should be large enough and require minimal effort to build.
- **Natural screens and walls**. Fallen trees, rock, large trees, etc. act as a heat reflector and aid in the dispersion of smoke from a fire.
- **Ventilation system**. The ventilation system provides fresh air, allows carbon monoxide to escape, reduces internal perspiration and acts as a heat sump (since hot air rises).
- **Pitch and tightness**. If the shelter is built with a parachute or other porous materials, remember that "pitch and tightness" apply to shelters designed to shed rain and snow. Porous materials will not shed moisture unless they are stretched tightly at an angle of sufficient pitch, which will encourage run-off instead of penetration. An angle of 40 to 60 degrees is recommended for the pitch of the shelter. The material stretched over the framework should be wrinkle-free and tight. Do not touch the material when water is running over it as this will break the surface tension at that point and allow water to drip into the shelter.
- **Insulation**. Insulation is added to shelters in colder environments in an effort to conserve the isolated person's loss of body heat due to conduction. Pine boughs, moss, and grasses can be used to insulate the shelter.
- **Framework**. Must be strong enough to support the weight of the covering and precipitation caused by snow or rain. It must also be sturdy enough to resist strong wind gusts.

IMMEDIATE ACTION SHELTER

6-13. An immediate action shelter is one which can be erected quickly with minimal effort. For example, from a raft, aircraft parts, parachutes, tarpaulin, or plastic material. Natural formations can also shield the isolated person from the elements immediately, to include overhanging ledges, fallen logs, caves, and tree wells (a tree well, also known as a "spruce trap," is the sheltered area under a tree that, in cold conditions, receives less snow than the surrounding area due to the overhanging branches, and consequently forms a pit in the snow[1]). Regardless of type, the shelter must provide whatever protection from the elements and enemy observation that is needed. In some instances, immediate action shelters may have to serve as permanent shelters. Examples of immediate-action shelters include—

- A wall or existing man-made structure.
- A vehicle or aircraft.
- A tarp, poncho, piece of plastic, space blanket, or evasion chart (EVC).
- Any material that is found in the area that will provide top cover.
- A cave or rock outcropping.

NATURALLY OCCURRING SHELTERS

6-14. Naturally occurring shelters should be considered whenever it is determined that shelter is necessary, especially when isolated persons are concerned with getting out of the elements and have little time, materials, or tools. Using what is already available prevents the need for the complete construction of a shelter from scratch, and will save, time, energy, and materials (see Figure 6-1 on page 197). Examples of naturally occurring shelters include—

- Caves.
- Rocky crevices.
- Small depressions.
- Large rocks on leeward sides of hills.
- Clumps of bushes.
- Large trees and low-hanging limbs.
- Fallen trees with thick branches.
- Uprooted tree buttresses.
- Tree pit snow shelters.

CAUTION: Particularly in harsh conditions, you may not be the only creature seeking shelter in natural formations. Always approach caves, crevices, etc. judiciously, with due consideration of the local animals you might encounter within.

---

1. Bear in mind that tree wells can be dangerously deep, meaning you could be unable to climb out. In experiments 90% of volunteers placed in tree wells were unable to free themselves without assistance. Subsequent snowfall may increase this hazard.

Figure 6-1. Types of Naturally Occurring Shelters

MANMADE SHELTER CONSTRUCTION

6-15. Shelters constructed from man-made material usually require more time, energy, and materials to construct than naturally occurring shelters. However, it is best to select a site that can be improved upon with minimal effort: always remember that economizing your effort reduces the quantity of food and water you will need to obtain. Existing trees can be used in various ways to cut down on the time and effort of constructing the framework. Isolated persons must gather the necessary materials to build the framework for some shelters. The man-made materials available dictate the type of shelter that can be constructed. Examples of man-made shelters include—

- Poncho lean-tos.
- Poncho tents.
- A-frames.
- Tepees.
- Para hammocks.

Poncho Lean-tos

6-16. It takes only a short time and minimal equipment to build this lean-to. For the lean-to, required items are a poncho, 7~10 feet of rope or parachute suspension line, three stakes about one foot long each, and two trees or two poles 7~10 feet apart. Before selecting the trees to be used or the location of the poles, check the wind direction. Ensure that the back of the lean-to will be into the wind (see Figure 6-2 on page 199). Use the following steps to construct a poncho lean-to:

Step 1 Tie off the hood of the poncho. Pull the drawstring tight, roll the hood long ways, fold it into thirds, and tie it off with the drawstring.

Step 2 Cut the rope in half. On one long side of the poncho, tie half of the rope to the corner grommet. Tie the other half to the other corner grommet.

Step 3 Attach a drip stick (about a 4-inch stick) to each rope about 1 inch from the grommet. These drip sticks will keep rainwater from running down the ropes into the lean-to. Tying strings (about 4 inches long) to each grommet along the poncho's top edge will allow the water to run to and down the line without dripping into the shelter.

Step 4 Tie the ropes about waist high on the trees. Use a round turn and two half hitches with a quick-release knot.

Step 5 Spread the poncho and anchor it to the ground, putting sharpened sticks through the grommets and into the ground.

Step 6 If the lean-to will be used for more than one night or rain is expected, make a center support for the lean-to. Make this support with a line. Attach one end of the line to the poncho hood and the other end to an overhanging branch. Make sure there is no slack in the line. Alternatively, place a stick upright under the center of the lean-to. However, this method will restrict space and movements in the shelter.

Step 7 For additional protection from wind and rain, place some brush, a rucksack, or other equipment at the sides of the lean-to.

Step 8 To reduce heat loss to the ground, place some type of insulating material, such as leaves or pine needles, inside the lean-to.

Note: When at rest, you can lose as much as 80 percent of your body heat to the ground.

Step 9 To increase security from enemy observation, lower the lean-to's silhouette by making two changes. First, secure the support lines to the trees at knee height (not at waist height) using two knee-high sticks in the two center grommets (sides of lean-to). Second, angle the poncho to the ground, securing it with sharpened sticks, as above.

Figure 6-2. Poncho Lean-to

Wind

Poncho Tent

6-17. This tent provides a low silhouette. It also protects isolated persons from the elements on two sides (see Figure 6-3 on page 200 and Figure 6-4 on page 200). However, it has less usable space and observation area than a lean-to, decreasing reaction time to enemy detection. To make this tent, you will need a poncho, two 5~8-foot-long ropes, six sharpened sticks about 1 foot long each, and two trees 7 to 10 feet apart. To construct a poncho tent, use the following steps:

Step 1 Tie off the poncho hood in the same way as the poncho lean-to.

Step 2 Tie a 5~8-foot-long rope to the center grommet on each side of the poncho.

Step 3 Tie the other ends of these ropes at about knee height to two trees 7 to 10 feet apart and stretch the poncho tight.

Step 4 Draw one side of the poncho tight and secure it to the ground pushing sharpened sticks through the grommets.

Step 5 Follow the same procedure on the other side.

Step 6 If a center support is needed, use the same methods as for the poncho lean-to. Another center support is an A-frame set outside but over the center of the tent. Use two 12~16-foot-long sticks, one with a forked end, to form the A-frame. Tie the hood's drawstring to the A-frame to support the center of the tent.

Figure 6-3. Poncho Tent Using Overhanging Branch

Figure 6-4. Poncho Tent Using Center Support

6-18. In open sandy terrain, soil can be placed into excess material, if available, to create pillars by stacking sand bags to elevate the outside and inside layers. The sides are left open to allow air to flow between layers (see Figure 6-5 on page 201).

Figure 6-5. Types of Man-made Shelters

18 in. above or below ground surface is preferred for coolest temperatures

12~18 in. between layers

A-Frame Shelters

6-19. Shelter construction begins with the framework, which must be strong enough to support the weight of the covering and precipitation (ice, snow, rain). It must also be sturdy enough to resist strong wind gusts and will typically have a 40-60 degree pitch for shedding precipitation and providing shelter room (see Figure 6-6 on page 203.)

6-20. A-frame shelters are typically built using the following materials:

- One 12~18-foot-long sturdy ridgepole with all projections cleaned off.
- Two bipod poles, approximately seven feet long each.

- Parachute material, normally five or six gores, or other natural or man-made material.
- Suspension lines or cord.
- Buttons or small objects placed behind gathers of material to provide a secure way of affixing suspension line to the material.
- Approximately 14 stakes, approximately 10 inches long each.

6-21. Once the materials are collected and readied for use, assemble the framework and then apply the fabric. The care and techniques used to apply the covering will determine the effectiveness of the shelter in shedding precipitation. If using parachute material, stretch the center seam tight, and then work from the back of the shelter to the front, alternating sides and securing the material to stakes or framework by using buttons and lines. Stretch the material tight by pulling the material 90 degrees to the wrinkles. If the material is not stretched tight, any moisture will pool in the wrinkles and leak into the shelter.

6-22. If natural materials are used for the covering, the shingle method should be used. Start at the bottom and work towards the top of the shelter, overlapping the bottom of each piece over the top of the preceding piece. This will allow water to drain off. Place the material on the shelter in sufficient quantity so that you cannot see through the shelter from the inside. Continue building the shelter as follows:

- Lash the two bipod poles together at eye-level height.
- Place the ridgepole, with the large end on the ground, into the bipod formed by the poles and secure with a square lash (see Appendix A for information on all knots, lashings, etc.).
- The bipod structure should be 90-degrees to the ridgepole and the bipod poles should be spread out to an approximate equilateral triangle with a 60-degree pitch. A piece of line can be used to measure this.
- Tie off approximately two feet of the apex in a knot and tuck this under the butt end of the ridgepole. Use half hitches and clove hitches to secure the material to the base of the pole.
- Place the center radial seam of the parachute piece or the center of the fabric on the ridgepole. After pulling the material taut, use half hitches and clove hitches to secure the fabric to the front of the ridgepole.
- Scribe or draw a line on the ground from the butt of the ridgepole to each of bipod poles. Stake the fabric down, starting at the rear of the shelter and alternately staking from side to side to the shelter front. Use a sufficient number of stakes to ensure the parachute material is wrinkle-free.
- Stakes should be slanted or inclined away from the direction of pull. When tying off with a clove hitch, the line should pass in front

of the stake first and then pass under itself to allow the button and line to be pulled 90-degrees to the wrinkle.

Figure 6-6. A-frame Shelters

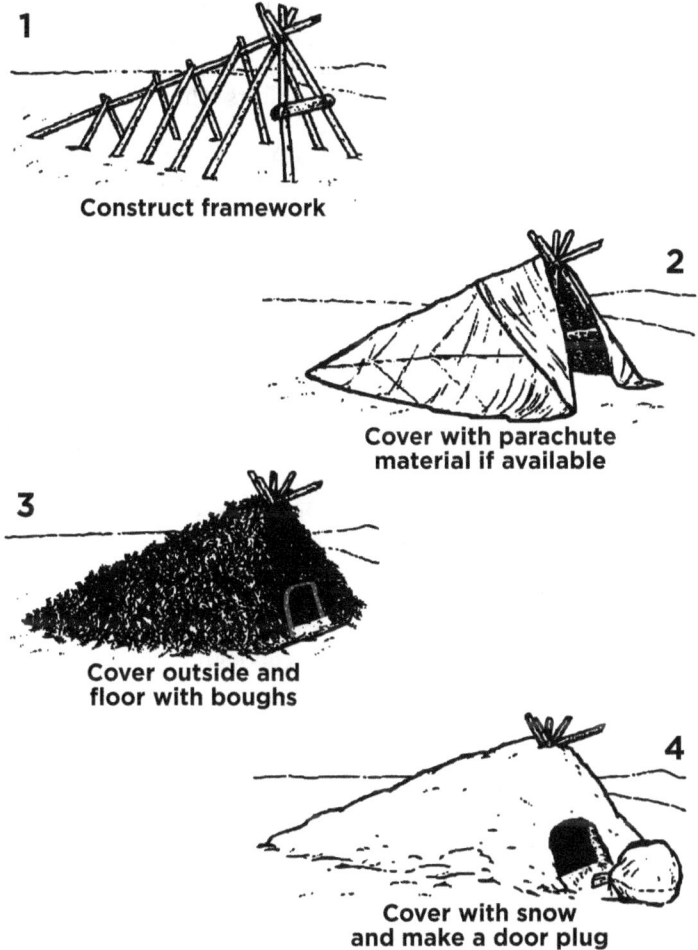

1

Construct framework

2

Cover with parachute material if available

3

Cover outside and floor with boughs

4

Cover with snow and make a door plug

NATURAL SHELTER CONSTRUCTION

6-23. Shelters can be constructed of natural materials such as existing trees, rocks, or other useful aspects found in the environment or terrain. These natural material shelters usually require the most time and energy to construct. It is extremely important to select a site that can minimalize expenditure of effort to ensure that the isolated person has the necessary energy. They may use a natural-made shelter when man-made shelter materials are limited or unavailable.

6-24. The environment or terrain will dictate what types of materials are available and ultimately what type of shelter isolated persons will construct. All natural materials should be collected before beginning construction of the shelter. This will include the framework, covering, bedding, insulation, and materials including lashing and staking for the shelter. If man-made lashing is unavailable, isolated persons may have to improvise cordage to be suitable. Branches may be securely placed against trees or other natural objects while support poles or branches can then be placed and/or attached depending on their function. They can use sticks with a Y fork for support or as a crossbeam. Some examples of natural made shelters are included in Paragraphs 6-25~6-38.

Snow Caves

6-25. The snow cave shelter (Figure 6-7 on page 205) is a most effective dwelling because of the insulating qualities of snow. Remember that it takes time and energy to build and that you will get wet while building it. First, you need to find a drift about three meters (10 feet) deep into which you can dig. While building this shelter, keep the roof arched for strength and to allow melted snow to drain down the sides.

6-26. Build the sleeping platform higher than the entrance. Separate the sleeping platform from the snow cave's walls or dig a small trench between the platform and the wall. This platform will prevent the melting snow from soaking you and your equipment. This construction is especially important if you have a good source of heat in the snow cave. Ensure the roof is high enough so that you can sit up on the sleeping platform.

6-27. Block the entrance with a snow block or other material and use the lower entrance area for cooking. The walls and ceiling should be at least 30 centimeters (1 foot) thick. Install a ventilation shaft. If you do not have a drift large enough to build a snow cave, you can make a variation of it by piling snow into a mound large enough to dig out.

Figure 6-7. Snow Cave

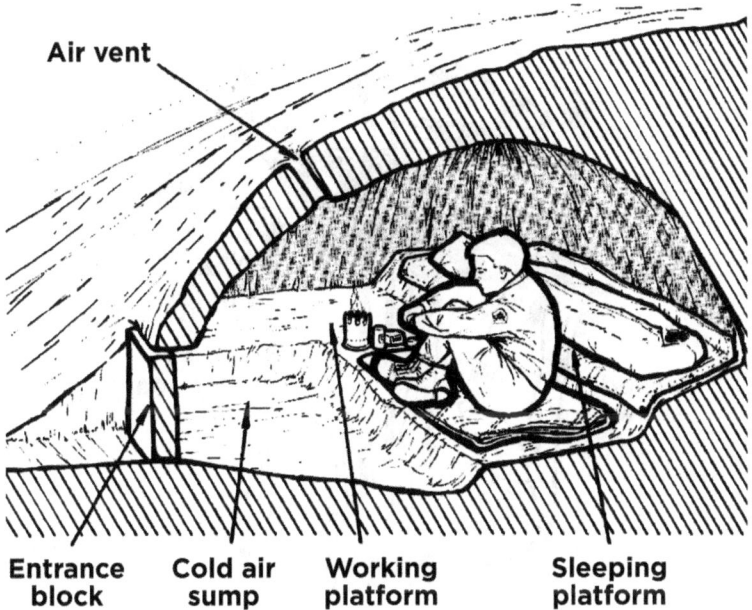

Air vent

Entrance
block

Cold air
sump

Working
platform

Sleeping
platform

Tree Pit Shelter

6-28. If you are in a cold, snow-covered area where evergreen trees grow and you have a digging tool, you can make a tree-pit shelter. (See Figure 6-8 on page 206). To make this shelter, you should—

- Find a tree with bushy branches that provides overhead cover.
- Dig out the snow around the tree trunk until you reach the depth and diameter you desire, or until you reach the ground.
- Pack the snow around the top and the inside of the hole to provide support.
- Find and cut other evergreen boughs. Place them over the top of the pit to give you additional overhead cover. Place evergreen boughs in the bottom of the pit for insulation.

Figure 6-8. Tree-Pit Snow Shelter

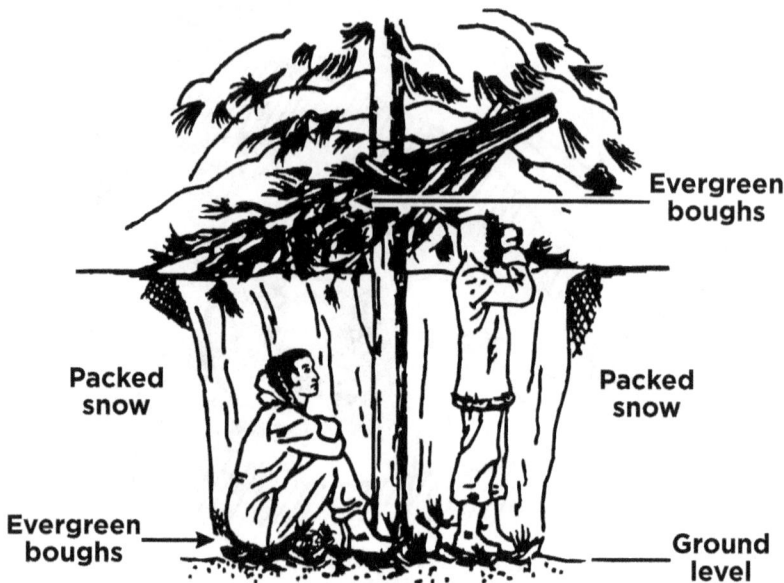

Raised Platform Shelter

6-29. This shelter type has many variations. One example is to use four trees or vertical poles in a rectangular pattern a little taller and a little wider than you are, keeping in mind you will also need protection for equipment. Square-lash two long, sturdy poles between the trees or vertical poles, one on each side of the intended shelter. Secure cross-pieces across the two horizontal poles at 6~12-inch intervals. This forms the platform on which a natural mattress may be constructed. Cover with an insect net of parachute material or other fabric. Figure 6-9 on page 207 shows an example of a raised platform shelter.

6-30. Construct a roof over the structure using A-frame building techniques. The roof should be waterproofed with thatching laid bottom to top in a thick shingle fashion. A variation of the platform-type shelter is the para platform. A quick and comfortable bed is made by simply wrapping material around the two "frame" poles. Another method is to roll poles in the material in the same manner as for an improvised stretcher.

Figure 6-9. Raised Platform Shelter

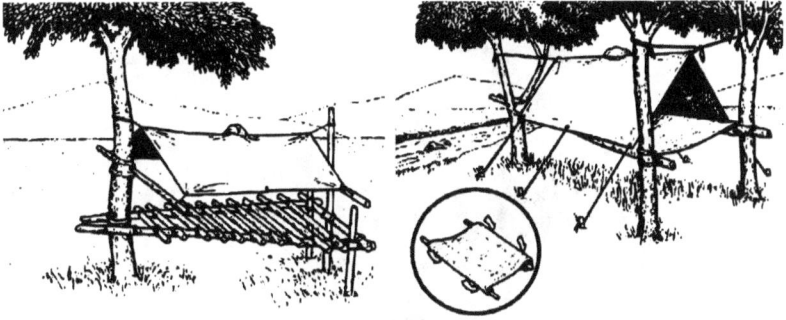

Field-expedient Lean-to

6-31. If you are in a wooded area and have enough natural materials, you can make a field-expedient lean-to (Figure 6-10 on page 208) without the aid of tools or with only a knife. It takes longer to make this type of shelter than it does to make other types, but it will protect you from the elements. To build this shelter, you will need—

- Two trees (or upright poles) about 2 meters (7 feet) apart.
- One pole about 2 meters (7 feet) long and 2.5 centimeters (1 inch) in diameter.
- Five to eight poles about 3 meters (10 feet) long and 2.5 centimeters (1 inch) in diameter for beams.
- Cord or vines for securing the horizontal support to the trees.
- Other poles, saplings, or vines to crisscross the beams.

6-32. To make this lean-to, you should—

- Tie the 2-meter (7-foot) pole to the two trees at waist to chest height. This is the horizontal support.

If a standing tree is not available, construct a bipod using Y-shaped sticks or two tripods.

- Place one end of the beams (3-meter (10-foot) poles) on one side of the horizontal support. As with all lean-to type shelters, be sure to place the lean-to's back side into the wind.
- Crisscross saplings or vines on the beams.
- Cover the framework with brush, leaves, pine needles, or grass, starting at the bottom and working your way up like shingling.
- Place straw, leaves, pine needles, or grass inside the shelter for bedding.

Figure 6-10. Field-expedient Lean-to and Fire Reflector

6-33. In cold weather, add to your lean-to's comfort by building a fire reflector wall. Drive four 1.5-meter-long (5-foot-long) stakes into the ground to support the wall. Stack green logs on top of one another between the support stakes. Form two rows of stacked logs to create an inner space within the wall that you can fill with dirt. This action not only strengthens the wall but also makes it more heat-reflective. Bind the top of the support stakes so that the green logs and dirt will stay in place.

6-34. With just a little more effort, you can have a drying rack. Cut a few 2-centimeter-diameter (3/4-inch- diameter) poles long enough to span the distance between the lean-to's horizontal support and the top of the fire reflector wall. Lay one end of the poles on the lean-to support and the other end on top of the reflector wall. Place and tie smaller sticks across these poles. You now have a place to dry clothes, meat, or fish.

Desert Shelters

6-35. More complex desert shelters can be built by using the same man-made materials and incorporating new features. However, any shelter that requires significant work should be done in the cooler parts of the day to reduce water loss and overheating. You can add multilayered roofs that create a middle airspace to capture and dissipate the sun's heat before it reaches the living area. A middle airspace of 12~18 inches can reduce the temperature in the shelter by 20~40 degrees Fahrenheit.

6-36. You can also dig out the bottom of the shelter or raise the bed from the ground surface with similar effect. Digging 18 inches below the desert surface can lower the temperature in the shelter by up to 40 degrees Fahrenheit. If available, light colored materials should be used on the outside to reflect heat, while the inside layer should be of dark color to filter out harmful ultraviolet rays. A below-ground desert shelter is an example that incorporates these features (see Figure 6-11).

Figure 6-11. Below-ground Desert Shelter

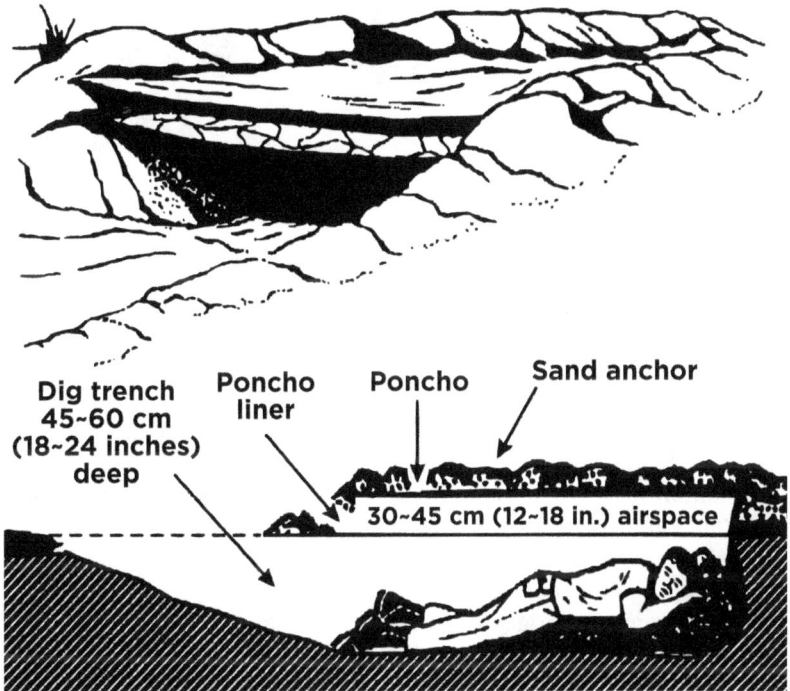

Dig trench 45~60 cm (18~24 inches) deep **Poncho liner** **Poncho** **Sand anchor**

30~45 cm (12~18 in.) airspace

Swamp Bed

6-37. The swamp bed is built in marshes or swamps or any area with standing water or wet ground (see Figure 6-12 on page 210).

- Look for four trees clustered in a rectangle or cut four poles and drive them in the ground (bamboo is ideal). The four corners should form a rectangle and should be far enough apart and strong enough to support your height and weight, including your equipment.

- Cut two poles that span the width of the rectangle. Secure the two poles to the post (or trees), making sure they are high enough to keep the bed above water and the ground to allow for high tide or high water.

- Cut additional poles, which span the rectangle's length, laying them across the two side poles and securing them. Cover the top of the bed frame with broad leaves or grass to form a softer sleeping surface. Build a fire pad by laying clay, silt, or mud at one corner of the swamp bed and allowing it to dry.

Figure 6-12. Swamp Bed

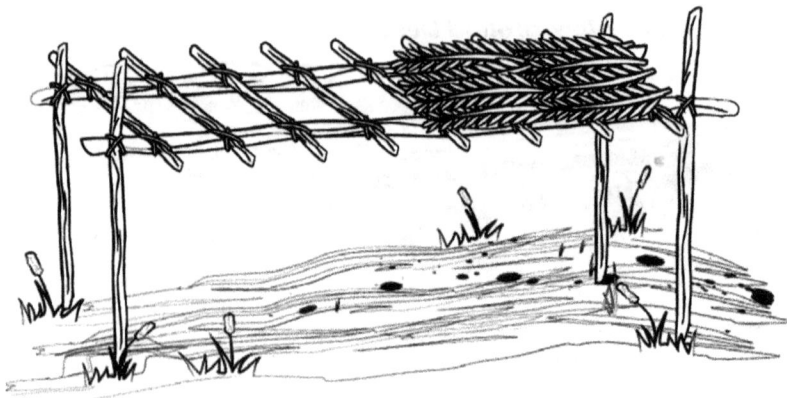

Debris Hut

6-38. For warmth and ease of construction, the debris hut is one of the best shelters available (see Figure 6-13 on page 211). When shelter is essential to survival, build this shelter. To make a debris hut, use the following steps:

Step 1 Make a tripod with two short stakes and a long ridgepole or by placing one end of a long ridgepole on top of a sturdy base.

Step 2 Secure the ridgepole (pole running the length of the shelter) using the tripod method or by anchoring it to a tree at about waist height.

Step 3 Prop large sticks along both sides of the ridgepole to create a wedge-shaped ribbing effect. Ensure that the ribbing is wide enough to accommodate the body and steep enough to shed moisture.

Step 4 Place finer sticks and brush crosswise on the ribbing. These form a latticework that will keep the insulating material (grass, pine needles, and leaves) from falling through the ribbing into the sleeping area.

Step 5 Add light, dry, soft debris over the ribbing until the insulating material is at least 3 feet thick—the thicker the better.

Step 6 Place a 2-foot layer of insulating material inside the shelter.

Step 7 At the entrance, pile insulating material that can be dragged to the entrance from inside the shelter to close the entrance or build a door.

Step 8 Finally, add shingling material or branches on top of the debris layer to prevent the insulating material from blowing away in a storm.

Figure 6-13. Debris Hut

THERMAL PRINCIPLES AND INSULATION

6-39. A minimum of 6~12 inches of insulation above the isolated person is needed to retain heat. All openings except ventilation holes should be sealed to avoid heat loss. Leaving vent holes open is especially important if heat-producing devices are used (see Figure 6-14 on page 212). Candles, fuel gel, and small oil lamps all produce carbon monoxide. In addition to the ventilation hole through the roof, another vent may be required at the door to ensure adequate circulation of air. As a rule, if you cannot see your breath, the snow shelter is too warm and should be cooled down to preclude melting and dripping.

6-40. If a fire is placed outside the shelter, the opening of the shelter should be placed 90 degrees to the prevailing wind. This reduces the chances of sparks and smoke being blown into the shelter if the wind reverses direction in the morning and evening, which frequently occurs in mountainous areas. The best fire-to-shelter distance is about 3 feet.

Figure 6-14. Placement of Ventilation Holes in a Snow Cave[1]

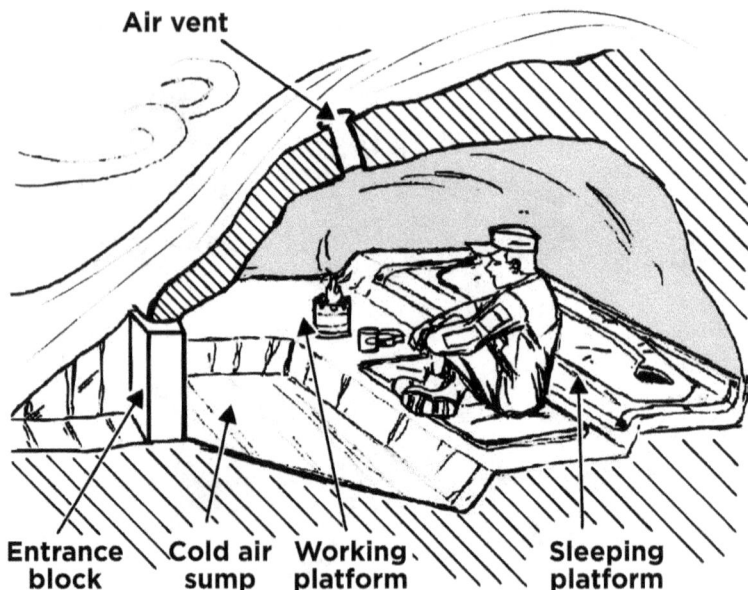

Air vent

Entrance block   Cold air sump   Working platform   Sleeping platform

6-41. It is unwise to build a fire in or near an aircraft or vehicle wreckage, especially if it is used as a shelter. The possibility of accidentally igniting spilled lubricants or fuels, with disastrous results, is high. Isolated persons may decide instead to use materials from the aircraft or vehicle to add to a shelter located a safe distance from the crash or vehicle abandonment site.

TIE-OFF POINTS

6-42. When working with a manufactured piece of shelter covering, such as a poncho or a tarp, you can secure the material in place using the grommets that are attached to the material. However, if using a piece of plastic or parachute material, no grommets will be available. In those instances, grab a small amount of material such as dirt, pine duff, an acorn, or other similar available items about 1 inch in diameter. Grasp the piece of shelter material and place the small amount of button material on the backside of the shelter material. Gather the material around the button material making a protrusion or round bulge. Put a small staking line with a slip loop over the round bulge in the shelter material and secure the slip loop so that the line is now temporarily attached to the material. The shelter material

---

1. Figure 6-14 is informationally identical to Figure 6-7 on page 205.

can now be staked to the ground and any wrinkles can be pulled out of the material to create the proper pitch and tightness.

STAKING

6-43. Staking the shelter to the ground is the preferred method of achieving a safe and secure foundation. To procure stakes, sharpen several sticks 1 inch in diameter by 12 inches in length to a point, so they can be easily driven into the ground at about a 45-degree angle to provide opposing force to the shelter cover. Secure the shelter cover to the stake with a piece of line using some type of hitch (the clove hitch works well for this task).

6-44. If the ground is rocky or frozen and the sharpened stake cannot be driven into the ground, place a large-diameter branch or rock on top on the ground near the shelter covering and secure the material to the branch. The weight on the branch will achieve the same goal of securing the shelter covering in place. If ground conditions are too soft, such as in loose sand or dirt, tie the shelter covering to a small piece of wood or a rock and bury it in a hole. Cover it up with dirt to hold it in place. This type of stake is known as the "dead man" stake or anchor.

BEDDING AND GROUND INSULATION

6-45. Heat loss through ground conduction, which carries a significant risk of hypothermia in cold environments, can be somewhat controlled by placing a layer of insulation between the isolated person and the cold ground. This can be achieved with a man-made sleeping pad made specifically for this purpose. Ground insulation can also be improvised by taking material from the environment like vehicle seat cushions, aircraft parts, and extra clothing. If these materials are limited, use natural materials from the environment such as ferns and grass (dry, if possible). In a coniferous forest, boughs from a tree will work.

6-46. When using natural material, make the insulation a minimum of 12 inches thick. This will allow sufficient insulation between isolated persons and the ground once the bed is compressed. The insulation should be fluffed in-between uses and material added to maintain comfort.

CHAPTER 7

MOVEMENT AND NAVIGATION

As an isolated person, your ability to walk effectively is important to your recovery as well as to your safety and the conservation of personal energy and resources. Equally important is the ability to perform land navigation: to determine where you are, where you wish to go, and how to get there efficiently. The ability to determine location and navigate cross-country significantly increases the isolated person's chances of recovery. Fundamental to land navigation is the acquisition of detailed knowledge about the specific operational environment to which you or your unit may be sent including climate, terrain, hydrology, topography, and prominent terrain features or landmarks. This chapter provides an overview on movement and navigation considerations, including the use of expedient navigation aids.

DECISION TO STAY OR MOVE

7-1. The decision to stay in place or move while isolated requires careful consideration. The reasons you may decide to stay in place or move are typically based upon the following:

- Requirements in your isolated Soldier guidance (ISG) or evasion plan of action (EPA) to move.
- Enemy situation. Do the enemy's composition, disposition or activity threaten your ability to evade or survive?
- If the current location does not provide multiple escape routes and/or contains numerous danger areas that canalize or funnel your activities and thus make them predictable to the enemy.
- If the current location does not provide adequate food, water, shelter, or the ability to report/communicate or support the recovery effort.

- Injury or circumstance that limits your ability (physical and mental) to move.
- An operational environment in which host-country military and law enforcement agencies have control as well as the intent and capability to assist recovery operations may favor staying near the isolation site (ex. aircraft crash) initially. Movement is considered only when certain that water, shelter, food, and help can be reached, or after having waited several days, if you are convinced that recovery is not coming and you are well-equipped to travel.
- Other factors that should be considered prior to deciding to move include the following:
  - Remember MARCH (see page 29). Do you have a head injury or other condition that affects your ability to make a clear decision? Avoid making any decision immediately after the isolating event. If possible, wait a period to allow for recovery from the mental (if not the physical) shock resulting from the event. When shock has subsided, clearly evaluate the situation, thoughtfully analyze the factors involved, and make valid decisions.
  - Are the necessary tools, equipment and materials available to support movement? Movement is risky unless the necessities of survival are available. To leave a safe shelter to travel in adverse weather conditions is foolhardy unless in an escape or evasion situation.

7-2. Once the decision is made to move, several considerations apply regardless of the circumstance:

- The ranking person must assume leadership and direct the efforts of the team during movement and recovery.
- Continually assess your physical capabilities (rest, hydration level, energy intake and output, injury prevention, pace and durability, mental and emotional state). Stop, think and take action *before* a problem arises.

MOVEMENT CONSIDERATIONS

7-3. Proper movement maximizes speed and mitigates energy output along a route. The best posture when moving should balance the person's body weight directly over their feet. Dependent upon the requirement to move stealthily, the soles of the feet should be flat on the ground. Step over obstacles and not directly on them as they may pose a serious tripping hazard. When ascending a terrain feature, lock your knees with each step to assist the joints in carrying your weight. When moving up steep slopes, use a zigzag motion ("traverse") to conserve energy and help maintain balance.

7-4. When descending a terrain feature, keep your back straight and keep your knees bent so you do not overextend the knee joints. The traverse is also used to descend steep terrain. After determining the step is stable, transfer weight from one foot to the other and repeat the process. Additional considerations include—

- Using game trails when they follow a projected course only in addition to an easier route of travel and the chance of securing game or locating water.
- Planning movement only after carefully surveying the surrounding countryside's terrain.
- Studying your back-trail carefully. Know your route backward equally as well as your route forward.
- Continually assessing the climate and geography as you move, asking yourself questions such as:
  - Is the weather changing?
  - Is the route leading you into an area that does not support food, water, or shelter?
  - Does the movement require a change in survival techniques?
- Making camp early so that you have plenty of time before nightfall to build a shelter and, if appropriate, a fire.
- Using the buddy system to watch for heat and cold injury.
- Not placing your hands or feet anywhere without first looking to see what is there.
- Observing grazing type animals (including deer) as they are better indicators of water than predators/carnivores.
- Bees seldom range more than four miles from their nests or hives. They will usually have a water source in this range. Ants need water. A column of ants marching up a tree is going to a small reservoir of trapped water.

Mountainous and Cold Movement

7-5. When conducting movement in a desert environment, you must consider the following:

- Avoid possible avalanches of earth, rock, and snow, as well as deep crevices in ice fields.
- Movement on the wind-packed side of a ridge is usually more advantageous because the snow surface is typically firmer and there is a better view of the route from above.
- A loose snow layer underneath a hard upper layer is more hazardous than a loose layer over a compacted one, as in the first case the upper layer of snow will slide more easily with no rough texture to restrain it. This is how "slab avalanches" (the most dangerous type of avalanche, responsible for the majority of avalanche fatalities) happen; vulnerability is caused by a cohesive "slab" layer of snow sitting on top of a weak layer.

- Leeward slopes collect snow that has been blown from the windward sides, forming slabs or "sluffs", depending upon the temperature and moisture. A sluff is a collection of dry, powdery, loose snow; a "sluff avalanche" is considerably less dangerous than a slab avalanche.
- Use a pole to probe ice and snow conditions during movement
- Avoid crossing glacial areas during the day.
- Avoid traveling during a blizzard.
- Take care when crossing thin ice, and do so only when absolutely necessary. Distribute your weight by lying flat and crawling.
- Cross streams where the water level is lowest.
- Wind-chill (the lowering of body temperature due to low-temperature air passing over the skin) is a factor in all activities. Personal movement generates wind-chill that can lead to increased cold weather injury.
- Have enough clothing to protect from the cold, and know how to maximize its warmth. For example, always keep your head covered. A person will lose 40 to 45 percent of their body heat from an unprotected head and even more from the unprotected neck, wrist, and ankles.
- The brain is very susceptible to cold and can stand the least amount of cooling of any body part. Because there is much blood circulation in the head, most of which is on the surface, you can lose heat very quickly if you do not cover your head.

Desert Movement

7-6. When conducting movement in a desert environment, you must consider the following:

- Avoid salt marshes. Water in these areas is typically undrinkable without significant purification effort. The area is highly corrosive to skin, equipment and clothing.
- Expect a large thermal shift between day and night. The drop in temperature at night occurs rapidly and will rapidly chill a person who lacks appropriate clothing.
- Protect radios and batteries from direct sunlight while moving in the desert environment.
- Rest during the day, work during the cool evenings and nights.
- Hide or seek shelter in dry washes (wadis) with thicker growths of vegetation and cover from oblique observation.
- Use the shadows cast from brush, rocks, or outcroppings. The air temperature in shaded areas will be cooler than that of exposed areas.
- Use the 1:3 rule when judging distance in the desert. What appears to be 1 kilometer away is really 3 kilometers away.

- Expect major sand and/or dust storms at least once a week. To avoid becoming lost, do not move during these storms.
- Mirage makes it difficult to identify targets, estimate range, and see objects clearly. Move to high ground (at least 10 feet or more above the desert floor) to get above the superheated air close to the ground and thus overcome the mirage effect.
- Find shade! Get out of the sun!
- Place something between you and the hot ground.
- If water is scarce, do not eat.

Jungle Movement

7-7. When conducting movement in a jungle environment, you must consider the following:

- There is less likelihood of recovery from beneath a dense jungle canopy than in other survival situations. Movement will be required.
- Avoid saltwater swamps if you can. If a saltwater swamp cannot be avoided and there are water channels through it, use a raft to cross it.
- Do not concentrate on the pattern of bushes and trees to your immediate front. Focus on the jungle further out and find natural breaks in the foliage. Look through the jungle, not at it. Stop and stoop down occasionally to look along the jungle floor.
- Move through the jungle. Do not fight the jungle. Turn your shoulders, shift your hips, bend your body, and shorten or lengthen your stride as necessary to slide between the undergrowth.
- Do not grasp at brush or vines when moving or use them to pull yourself up when climbing slopes; they may have irritating spines or sharp thorns.
- Protect yourself from insects with netting, clothing, etc.
- Promptly treat wounds and scratches, no matter how minor, to avoid dangerous infection.

WATER CROSSINGS

7-8. A huge variety of adjectival description can be applied to rivers and streams. They may be shallow or deep, slow or fast moving, narrow or wide. Before crossing a river or stream, develop a good plan. The first step is to look for a high place from which to get a good overview of the river or stream. From here, you can look for a place to cross.

Rivers and Streams

7-9. Good crossing locations include—

- A level stretch where it breaks into several channels. Two or three narrow channels are usually easier to cross than a single wide river.

- A shallow bank or sandbar. If possible, select a point upstream from the bank or sandbar so that the current will carry you to it if you lose your footing.
- A course across the river that leads downstream so that you will cross the current at about a 45- degree angle.
- Remember that each part of the river must necessarily carry the same volume of water over a given period of time as every other part (except where the volume is added to by tributaries). Therefore you can expect a widening of a river to indicate shallower and/or slower water, and a narrowing of a river to indicate deeper and/or faster water.

7-10. The following areas possess potential hazards; avoid them if possible:

- A ledge of rocks that crosses the river. This often indicates dangerous rapids or canyons.
- An estuary of a river, because it is normally wide, has strong currents and is subject to tides. These tides can influence some rivers many miles from their mouths; you should go back upstream to an easier crossing site.
- Eddies, which can produce a powerful backward pull downstream of the obstruction causing the eddy and pull you under the surface.

Rapids

7-11. To swim across a deep, swift river, swim with the current; do not swim against the current. Try to keep your body horizontal to the water: this will reduce the danger of being pulled under. In fast, shallow rapids, lie on your back, feet pointing downstream, finning your hands alongside your hips. This action will increase buoyancy and help steer away from obstacles. Keep your feet up to avoid getting them caught by rocks.

7-12. In deep rapids, lie on your stomach, head downstream, and angle toward the shore. Watch for obstacles and be careful of backwater eddies and converging currents, as they often contain dangerous swirls. Converging currents occur where new watercourses enter the river or where water has been diverted around large obstacles such as small islands. To ford a swift, treacherous stream, apply the following steps:

Step 1 Find a strong pole about 3 inches in diameter and 7~8 feet long to help ford the stream.

Step 2 Grasp the pole and plant it firmly on the upstream side to break the current. Plant feet firmly with each step, and move the pole forward a little downstream from its previous position, but still upstream.

Step 3 With the next step, place foot below the pole. Keep the pole well slanted so that the force of the current keeps the pole against your shoulder. Cross the stream in such a manner that the downstream current is being crossed at a 45-degree angle, (see Figure 7-1).

Figure 7-1. Using a Pole Method to Ford Stream

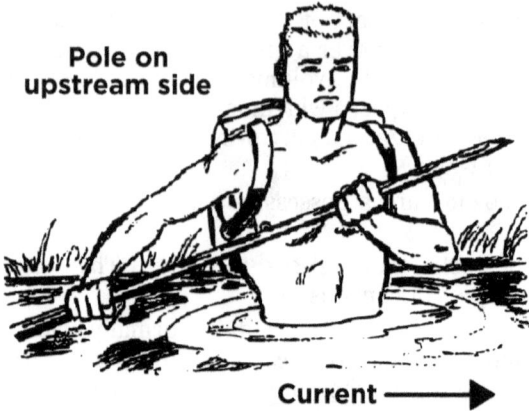

Pole on
upstream side

Current ———▶

7-13. If there are other people with you, ensure that they all cross the stream together. Position the heaviest person on the downstream end of the pole and the lightest person on the upstream end (see Figure 7-2 on page 221). In using this method, the upstream person breaks the current, and those below can move with relative ease in the eddy formed by the upstream person. If the upstream person temporarily loses footing, the others can hold steady while the upstream person regains footing. If there are three or more people crossing the stream and a rope is available, the technique shown in Figure 7-3 on page 222 can be used. The length of the rope must be three times the width of the stream.

Figure 7-2. Using a Pole for Multiple Crossing

Lightest man in upstream position

Heaviest man acts as downstream anchor for crossing

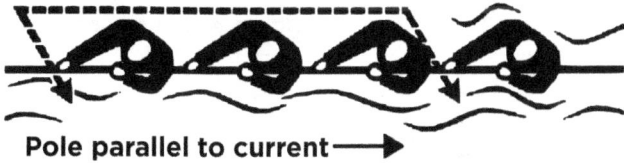

Pole parallel to current ⟶

Figure 7-3. River Crossing Using Rope

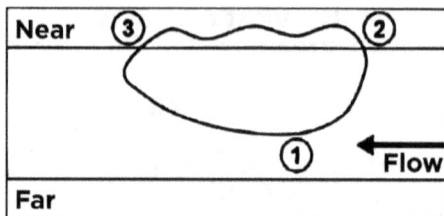

The person crossing is secured to the loop around the chest. The strongest person crosses first. The other two are not tied on— they pay out the rope as it is needed and can stop the person crossing from being washed away.

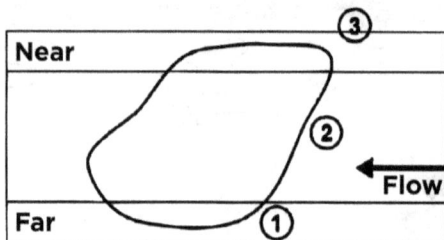

When he reaches the far bank, number 1 unties himself and number 2 ties on, number 2 crosses, controlled by the others. Any number of people can be sent across this way.

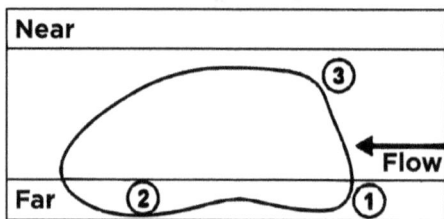

When number 2 has reached the far bank, number 3 ties on and crosses. Number 1 takes most of the strain, but number 2 is ready in case anything goes wrong.

Rafts

7-14. Rafts are a useful piece of equipment that enables a safer water crossing when you are required to cross a large body of water, cross with equipment, or your swimming skills are hindered by injury. If two ponchos are available, construct a brush raft or an Australian poncho raft. Using either of these rafts, equipment can be safely floated across a slow-moving stream or river. The following information will explain how to build field-expedient rafts.

Brush Raft

7-15. The brush raft is constructed out of ponchos, fresh green brush, two small saplings, and rope or vine as follows (see Figure 7-4 on page 223):

Step 1 Push the hood of each poncho to the inner side and tightly tie off the necks using the drawstrings.

Step 2 Attach the ropes or vines at the corner and side grommets of each poncho. Make sure they are long enough to cross to and tie with the others attached at the opposite corner or side.

Step 3 Spread one poncho on the ground with the inner side up. Pile fresh, green brush (no thick branches) on the poncho until the brush stack is about 18 inches high. Pull the drawstring up through the center of the brush stack.

Step 4 Make an X-frame from two small saplings and place it on top of the brush stack. Tie the X-frame securely in place with the poncho drawstring.

Step 5 Pile another 18 inches of brush on top of the X-frame, and then compress the brush slightly.

Step 6 Pull the poncho sides up around the brush and, using the ropes or vines attached to the corner or side grommets, tie them diagonally from corner to corner and from side to side.

Step 7 Spread the second poncho, inner side up, next to the brush bundle.

Step 8 Roll the brush bundle onto the second poncho so that the tied side is down. Tie the second poncho around the brush bundle in the same manner as the first poncho was tied around the brush.

Step 9 Place the raft in the water with the tied side of the second poncho facing up.

Figure 7-4. Brush Raft

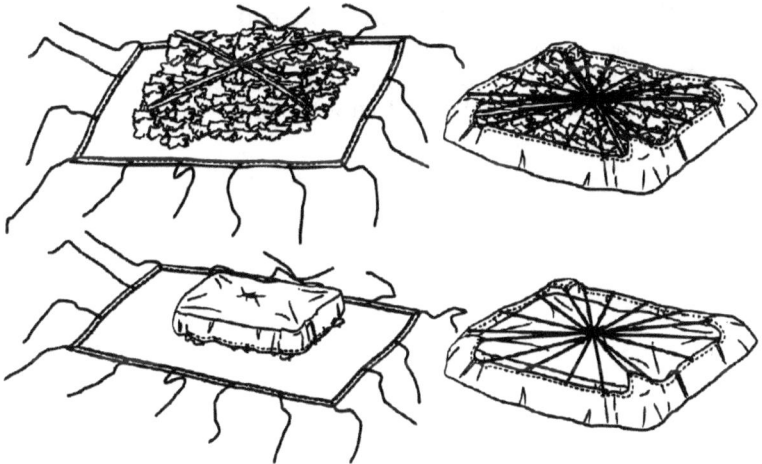

Australian Poncho Raft

7-16. The Australian poncho raft is constructed when there is no time to gather brush for a brush raft. This raft, although more waterproof than the poncho brush raft, will only float about 77 pounds of equipment. To construct this raft, use two ponchos, two rucksacks, two 4-foot poles or branches, and ropes, vines, bootlaces, or comparable material as follows (see Figure 7-5 on page 225):

Step 1 Push the hood of each poncho to the inner side and tightly tie off the necks using the drawstrings.

Step 2 Spread one poncho on the ground with the inner side up. Place and center the two 4-foot poles on the poncho about 18 inches apart.

Step 3 Place rucksacks, packs, or other equipment between the poles. Also, place other items that need to be kept dry between the poles. Snap the poncho sides together.

Step 4 Use a friend's help to complete the raft. Hold the snapped portion of the poncho in the air and roll it tightly down to the equipment. Make sure to roll the full width of the poncho.

Step 5 Twist the ends of the roll to form pigtails in opposite directions. Fold the pigtails over the bundle and tie them securely in place using ropes, bootlaces, or vines.

Step 6 Spread the second poncho on the ground, inner side up. If more buoyancy is needed, place some fresh green brush on this poncho.

Step 7 Place the equipment bundle, tied side down, on the center of the second poncho. Wrap the second poncho around the equipment bundle following the same procedure as used for wrapping the equipment in the first poncho.

Step 8 Tie ropes, bootlaces, vines, or other binding material around the raft about 12 inches from the end of each pigtail. Place and secure weapons on top of the raft.

Step 9 Tie one end of a rope to an empty canteen and the other end to the raft. This will help to tow the raft.

Figure 7-5. Australian Poncho Raft

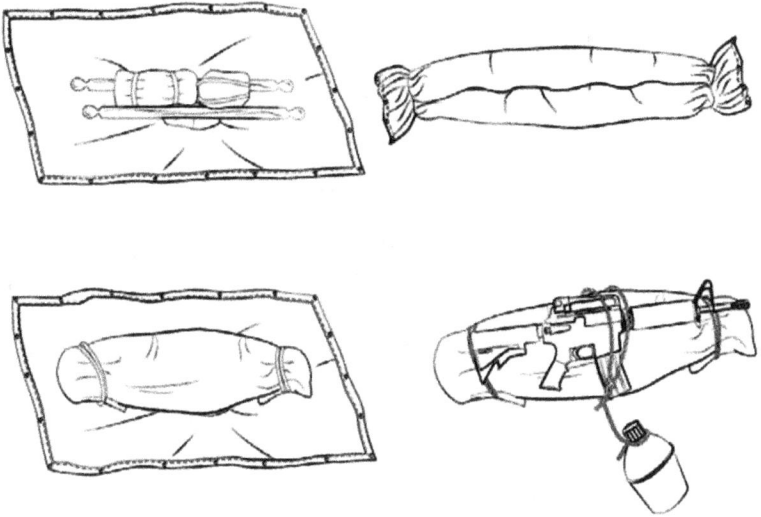

7-17. When launching any of the above rafts, take care not to puncture or tear it by dragging it on the ground. Before starting to cross a river or stream, let the raft lay on the water a few minutes to ensure that it floats. If the river is too deep to ford, you should push the raft in front of you while you are swimming. The design of the above rafts does not allow them to carry a person's full body weight; they should be used as a float to move you and your equipment safely across a river or stream.

Log Raft

7-18. The log raft is constructed using dry, dead, or standing trees for logs. Spruce trees found in polar and subpolar regions make the best rafts. A simple method for making a raft is to use pressure bars lashed securely at each end of the raft to hold the logs together (see Figure 7-6 on page 226).

Figure 7-6. Log Raft

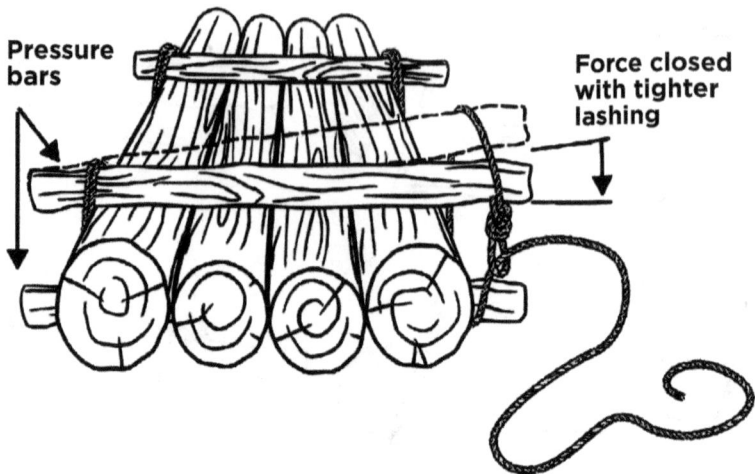

Pressure bars

Force closed with tighter lashing

Flotation Devices

7-19. If the water is warm enough for swimming and you do not have the time or materials to construct a poncho-type raft, you can use various flotation devices to negotiate the water obstacle. Items that can be used for flotation devices include—

- **Trousers**. Knot each trouser leg at the bottom and close the fly. With both hands, grasp the waistband at the sides and swing the trousers in the air to trap air in each leg. Quickly press the sides of the waistband together and hold it underwater so that the air will not escape. You now have water wings to keep you afloat as you cross the body of water.

Note: Wet the trousers before inflating to trap the air better; this may need to be done several times when crossing a large body of water.

- **Empty containers**. Lash together empty gas cans, water jugs, ammunition cans, boxes, or other items that will trap or hold air. Use them as water wings. Use this type of flotation device only in a slow-moving river or stream.

- **Plastic bags and ponchos**. Fill two or more plastic bags with air and secure them together at the opening. Use a poncho and roll green vegetation tightly inside it to form a roll at least 8 inches in diameter. Tie the ends of the roll securely. This can be worn around the waist or across one shoulder and under the opposite arm.

- **Logs**. Use a stranded drift log, if one is available, or find a log near the water to use as a float. Test the log before starting to cross. Some tree logs, such as palm, will sink even when the wood is dead. Another method is to tie two logs about 24 inches apart and sit between the logs with your back against one and your legs over the other (see Figure 7-7).

- **Cattails**. Gather stalks of cattails (tall, reed-like marsh plants characterized by a large, cylindrical, sausage-shaped seed head section near the tip of their stalks) and tie them in a bundle at least 10 inches in diameter. The many air cells in each stalk cause a stalk to float until it rots. You should test the cattail bundle to be sure it will support your weight before trying to cross a body of water.

Figure 7-7. Two-log Raft

7-20. There are many other flotation devices that can be devised by using some imagination. Just make sure to test the device before trying to use it.

DETERMINE CARDINAL DIRECTION

7-21. Using the sun and shadows. The earth's relationship to the sun can help you to determine direction on earth. The sun always rises in the east and sets in the west, but not exactly due east or due west. There is also some seasonal variation. Shadows will move in the opposite direction of the sun. In the Northern Hemisphere, they will move from west to east, and will point north at noon. In the Southern Hemisphere, shadows will indicate south at noon. With practice, you can use shadows to determine both direction and time of day.

SHADOW TIP METHOD

7-22. The shadow-tip method is ineffective for use north of the Arctic Circle or south of the Antarctic Circle due to the position of the sun above the horizon. Whether the sun is north or south of you at mid-day depends on your latitude. North of the Tropic of Cancer, the sun is always due south at local noon and the shadow points north. South of the Tropic of Capricorn, the sun is always due north at local noon and the shadow points south. In the tropics, the sun can be either north or south at noon, depending on the date and location, but the shadow progresses to the east regardless of the date. This method consists of four basic steps (see Figure 7-8):

Step 1 Place a stick or branch into the ground at a level spot where a distinct shadow will be cast. Mark the shadow tip with a stone, twig, or other means. This first shadow mark is always the westerly direction. (Note. The sun "rises in the east and sets in the west" (but rarely due east and due west). The shadow tip moves in just the opposite direction. Therefore, the first shadow-tip mark is always in the west direction, and the second mark is always in the east direction, any place on earth.

Step 2 Wait 10~15 minutes until the shadow tip moves a few inches. Mark the new position of the shadow tip in the same way as the first.

Step 3 Draw a straight line through the two marks to obtain an approximate east-west line.

Step 4 Standing with the first mark (west) to your left, the other directions are simple; north is to the front, east is to the right, and south is to the rear.

Note: A line drawn at right angles to the east-west line at any point forms the same approximate north-south line, which will help orient a person to the same cardinal directions.

Figure 7-8. Shadow-tip Method

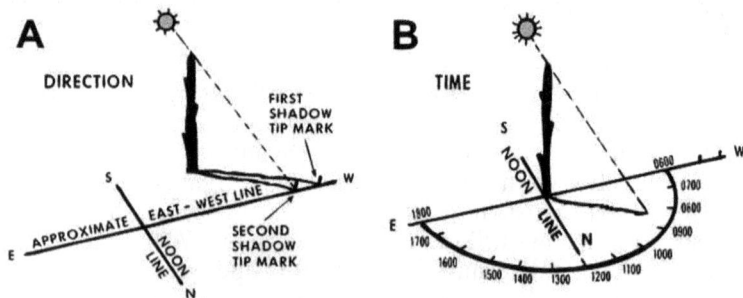

7-23. Inclining the stick to obtain a more convenient shadow does not impair the accuracy of the shadow-tip method. Therefore, an isolated person on sloping ground or in highly vegetated terrain need not waste valuable time looking for a large, level area. A flat spot, the size of the hand, is all that is necessary for shadow-tip markings and the base of the stick can be either above, below, or to one side of it.

7-24. In addition, any stationary object (the end of a tree limb or the notch where branches are jointed) serves just as well as an implanted stick because only the shadow tip is marked. The shadow-tip method can also be used to find the approximate time of day as follows:

- Move the stick to the intersection of the east-west line and the north-south line, and set it vertically in the ground.
- The west part of the east-west line indicates the time is 0600 and the east part is 1800.
- The north-south line now becomes the noon line. The shadow of the stick is the hour hand in the shadow clock and with it – estimate time by using the noon line and the six o'clock line as the guides. Depending on the location and the season, the shadow may move either clockwise or counterclockwise, but this does not alter the manner of reading the shadow clock.
- The shadow clock is not a timepiece in the ordinary sense. It always reads 0600 at sunrise and 1800 at sunset. However, it does provide a satisfactory means of telling time in the absence of properly set watches. Being able to establish the time of day is important for such purposes as keeping a rendezvous, prearranged concerted action by separated persons or groups, and estimating the remaining duration of daylight. Shadow-clock time is closest to conventional clock time at midday, but the spacing of the other hours, compared to conventional time, varies somewhat with the locality and date.

EQUAL-SHADOW METHOD

7-25. This method determines direction and is a variation of the shadow-tip method, (see Figure 7-9 on page 230). It is more accurate and may be used at all latitudes less than 66° (i.e., outside the Arctic Circle and Antarctic Circle) at all times of the year. It consists of the following four steps:

Step 1 Place a stick or branch into the ground vertically at a level spot where a shadow at least 12 inches long will be cast. Mark the shadow tip with a stone, twig, or other means. Mark your first shadow in the morning.

Step 2 Trace a circle around the stick using the shadow as the circumference and the base of the stick as the center, using a piece of string, shoelace, or a second stick.

Step 3 As noon approaches, the shadow becomes shorter and may disappear. After noon, the shadow lengthens until it crosses the circumference of the circle you drew earlier. Mark the spot as soon as the shadow tip touches the circle a second time.

Step 4 Draw a straight line through the two marks to obtain an accurate east-west line.

Figure 7-9. Equal-shadow Method

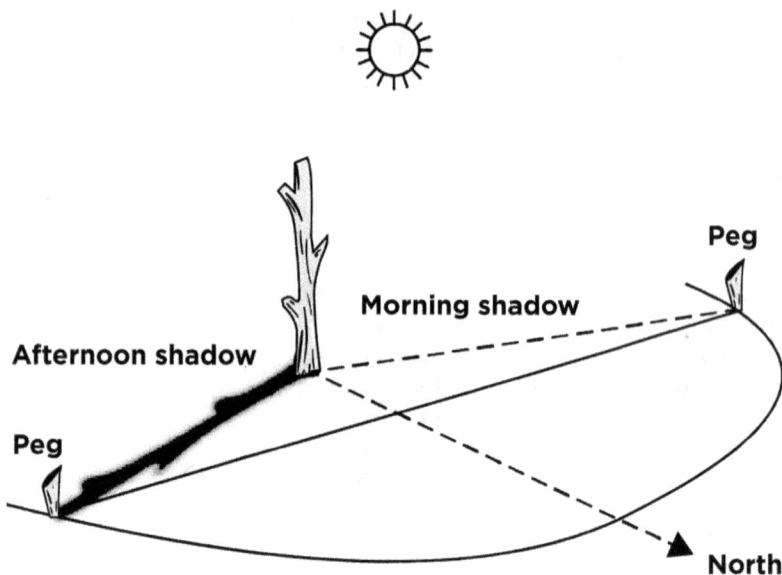

Peg

Morning shadow

Afternoon shadow

Peg

North

WATCH METHOD

7-26. This method requires the use of a common analog watch—one that has hands (if you have only a digital watch, see Paragraph 7-29). Use of this type of watch will enable the isolated person to determine approximate true north and true south (see Figure 7-10 on page 231) if it is set accurately to local time. In the North Temperate Zone, point the hour hand towards the sun. A south line can be found midway between the hour hand and the 12 o'clock marker on the watch dial, standard time. If on daylight savings time, the north-south line is found between the hour hand and the 1 o'clock marker. If there is any doubt as to which end of the line is north, remember that the sun is in the east before noon and in the west after noon.

7-27. The watch may also be used to determine direction in the South Temperate Zone. However, the method is different. The 12 o'clock marker on the watch dial is pointed toward the sun, and halfway between the 12 o'clock marker and the hour hand will be a north line.

If on daylight savings time, the north line lies midway between the hour hand and the 1 o'clock marker on the watch dial.

7-28. The watch method can be inaccurate, especially in the lower latitudes (i.e., nearer the equator), and may cause circling. Use with caution, or choose a different direction-finding method if possible.

7-29. If you have only a digital watch, draw a clock face on a circle of paper with the correct time on it and use it to determine your direction at that time. Even if you have an analog watch, you may also choose to draw a large clock face on the ground or lay your watch on the ground for a more accurate reading.

Figure 7-10. Watch Method

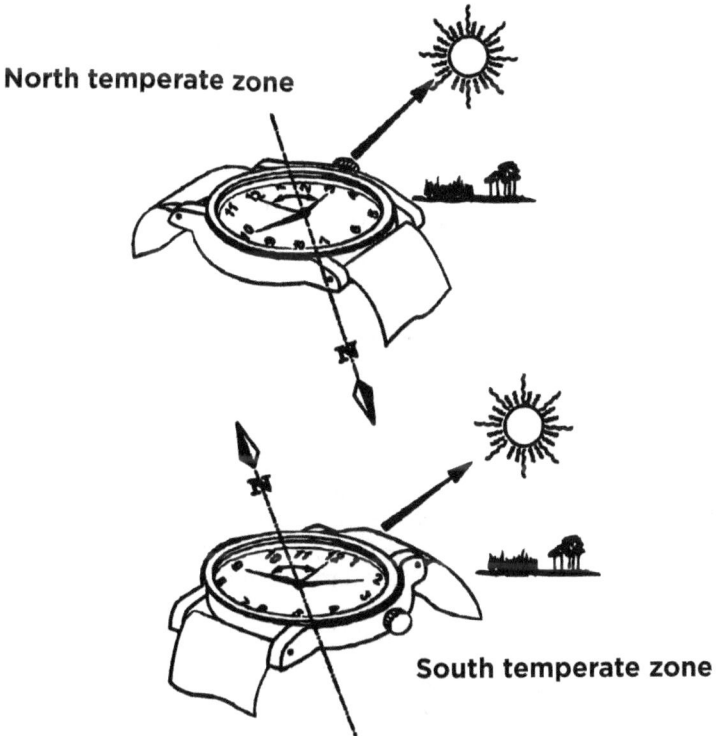

North temperate zone

South temperate zone

24-HOUR CLOCK METHOD

7-30. To utilize this method, take the local military time and divide it by two. In the Northern Hemisphere, point the relevant hour marker on your watch dial at the sun, and the 12 o'clock marker will point north. For example, it is 1400 hours. Divide 1400 by two and the answer is 700, which represents the hour. Holding the watch horizontal, point the 7 o'clock marker at the sun and 12 o'clock marker will point north. In the Southern Hemisphere, point the 12 o'clock marker at the sun, and the resulting "hour" from the division will point south.

Note: Bear in mind that this method does not require a functioning analog watch, requiring only that you know the current local time. A watch-face drawn on a piece of paper or inscribed on the ground is just as good (with practice, you can use this method without any physical aids at all).

USING THE MOON

7-31. Since the moon has no light of its own, we can only see it when it reflects the sun's light. As it orbits the earth on its 28-day circuit, the shape of the reflected light varies according to its position. We say there is a new moon or no moon when it is on the opposite side of the Earth from the sun. Then, as it moves away from the earth's shadow, it begins to reflect light from its right side and waxes to become a full moon before waning, or losing shape, to appear as a sliver on the left side. If the moon rises before the sun has set, the illuminated side will be the west. If the moon rises after midnight, the illuminated side will be the east. This obvious discovery provides a rough east-west reference during the night.

USING THE STARS

7-32. The choice of constellation to be used for determining north or south direction depends on whether you are in the Northern or Southern Hemisphere. Each sky is explained as follows:

Northern Sky

7-33. The main constellations to learn are Ursa Major (also known as the Big Dipper, the Great Bear, or the Plow), Ursa Minor (also known as the Little Dipper or the Little Bear), and Cassiopeia (also known as the Lazy W). Use them to locate Polaris, also known as the Pole star or the North Star (see Figure 7-11 on page 233). Polaris is considered to remain stationary, as it rotates only 1.08 degrees around the northern celestial pole. The North Star is the last star of the Little Dipper's handle and can be confused with the Big Dipper. However, the Little Dipper is made up of seven rather dim stars and is not easily seen unless far away from any town or city lights. Confusion can be prevented by using both the Big Dipper and Cassiopeia together. The Big Dipper and Cassiopeia are generally opposite each other and rotate counterclockwise around Polaris, with Polaris in the center. The Big Dipper is a seven-star constellation in the shape of a dipper (a ladle or scoop).

Stars

7-34. Forming the outer lip of the Big Dipper are the two **"pointer stars"**—known as such because they point to the North Star. Mentally draw a line from the outer bottom star to the outer top star of the Big Dipper's bucket. Extend this line about five times the distance between the pointer stars. The North Star will be located along this line.

7-35. The North Star can always be found at the same approximate vertical angle above the horizon as the northern line of latitude of the current location. For example, if at 35 degrees north latitude, Polaris will be easier to find if you scan the sky at 35 degrees off the horizon. This will help to lessen the area of the sky in which to locate the Big Dipper, Cassiopeia, and the North Star.

7-36. Cassiopeia (or the Lazy W) has five stars that form a shape like a "W." One side of the "W" appears flattened or "lazy." The North Star can be found by bisecting the angle formed by the three stars of the lazy side. Extend this line about five times the distance between the bottom of the "W" and the top. The North Star is located between Cassiopeia and Ursa Major (the Big Dipper). After locating the North Star, locate the North Pole or true north by drawing an imaginary line directly to the earth (see Figure 7-11).

Figure 7-11. Northern Sky

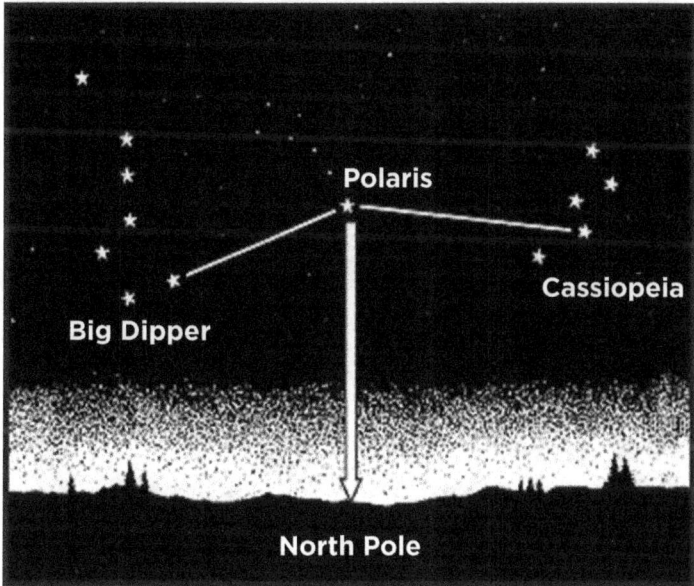

Southern Sky

7-37. There is no single star bright enough to clearly identify the south celestial pole. Therefore, use a constellation known as the Southern Cross (also known as Crux). You can use it as a signpost to the south. The Southern Cross has five stars. Its four brightest stars form a cross. The two stars that make up the cross's long axis are used as a guideline.

7-38. To determine south, imagine a distance four and one-half to five times the distance between these stars and the horizon. The pointer stars to the left of the Southern Cross serve two purposes. First, they provide an additional cue toward south by imagining a line from the stars toward the ground. Second, the pointer stars help accurately identify the true Southern Cross from the False Cross[1]. The intersection of the Southern Cross and the two pointer stars is very dark and devoid of stars. This area is called the Coalsack Nebula (also known as the Southern Coalsack, or simply the Coalsack). Look down to the horizon from this imaginary point to identify the South Pole and select a landmark to steer by (see Figure 7-12). In a static survival situation, isolated persons can fix this location for use in daylight by driving stakes into the ground at night to point the way.

Figure 7-12. Southern Sky

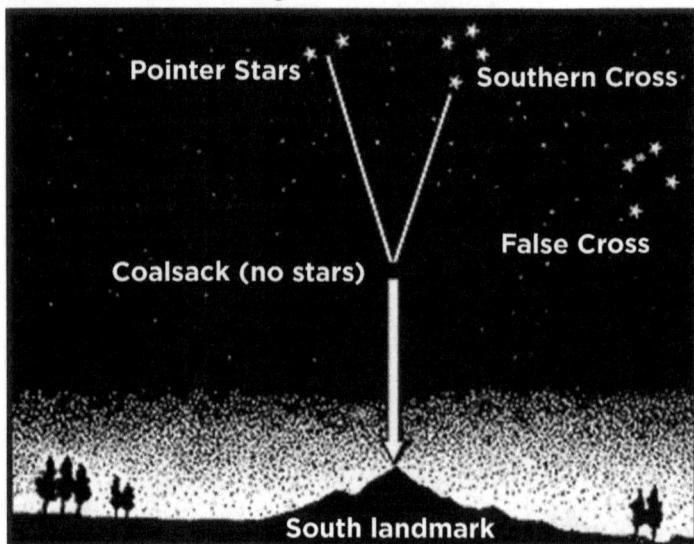

---

1. The False Cross is dimmer and larger than the true Southern Cross, but is still easily mistaken: ensure you can see the pointer stars that identify the correct constellation.

7-39. Depending on the star selected for navigation, azimuth checks (actions to verify your direction of travel) are necessary. A star near the north horizon serves for about half an hour. When moving south, azimuth checks should be made every 15 minutes. When moving east or west, the difficulty of staying on azimuth is caused more by the likelihood of the star climbing too high in the sky or losing itself behind the western horizon than it is by the star changing direction angle. When this happens, it is necessary to change to another guide star.

IMPROVISED COMPASS

7-40. An improvised compass can be constructed using a piece of ferrous (iron-containing) metal,[1] either needle-shaped or a flat double-edged razor blade, and a piece of thread or long hair from which to suspend it. Magnetize or polarize the metal by slowly stroking it in one direction on a piece of silk or carefully through your hair using deliberate strokes. You can also polarize the metal by stroking it repeatedly at one end with a magnet. Always stroke in one direction only. Suspend the needle or blade with the hair. The needle will gradually align itself in the north-south direction.

7-41. If you have a battery and some electrical wire, you can polarize the metal electrically. The wire should be insulated. If it is not insulated, wrap the metal object in a single, thin strip of paper or a leaf to prevent contact. The battery must be a minimum of 2 volts. Remove a small amount of insulation from each end of the wire, so you can make an electrical connection with the battery later. Wind the wire around the metal object to form a tight coil, avoiding electrical contact between the wire and the metal object. Ensure you leave enough of the wire protruding from the coil to reach the battery's positive and negative terminals. Touch the ends of the wire to the battery's terminals for a few seconds (one end touching the positive terminal, the other end touching the negative terminal). The metal object will become magnetic. When suspended from a piece of nonmetallic string or floated on a small piece of wood, cork, or a leaf in water (see Paragraph 7-43), it will align itself with a north-south line.

7-42. You can construct a more elaborate improvised compass using a sewing needle or other thin metallic object, a nonmetallic container (for example, the cut-off bottom of a plastic container or soft drink bottle), and the tip of a ballpoint or similar pen. To construct this compass, take an ordinary sewing needle and break it in half. One half will form the direction pointer and the other will act as the pivot point. Push the portion used as the pivot point through the bottom

---

1. Metals that **cannot** easily be magnetized include aluminum, gold, silver and copper.

center of the container; this portion should be flush on the bottom and not interfere with the lid. Attach the center of the other portion (the pointer) of the needle on the pen's tip using glue, tree sap, or melted plastic. Magnetize one end of the pointer and rest it on the pivot point.

FLOATING NEEDLE OR LEAF COMPASS

7-43. Magnetize a metal sewing needle by stroking it in one direction with a piece of silk or by running it across a small magnet. Oil the needle by passing it through your hair. Place the needle gently on the water surface. The oil on the needle will cause the needle to float on the water surface. The needle will gradually align itself in the north-south direction. As an alternative, the magnetized needle can be placed on a small piece of cork, leaf, duct tape, etc. The needle will float and will gradually align itself in the north-south direction.

OTHER MEANS OF DETERMINING DIRECTION

7-44. Moss growing on a tree cannot be relied on to indicate north because moss grows completely around some trees. Actually, growth is lusher on the south-facing side of trees in the northern hemisphere and on the north-facing side of trees in the southern hemisphere. If there are several felled trees around for comparison, look at the stumps. Growth is more vigorous on the side toward the equator, and the tree growth rings will be more widely spaced. Conversely, the tree growth rings will be closer together on the side toward the poles.

7-45. Wind direction may be helpful in some instances where there are prevailing wind directions and you know what those directions are.

7-46. Recognizing the differences between vegetation and moisture patterns on north- and south-facing slopes can aid in determining direction. In the northern hemisphere, north-facing slopes receive less sun than south-facing slopes and are therefore cooler and damper. In the summer, north-facing slopes retain patches of snow. In the winter, trees and open areas on south-facing slopes and the southern side of boulders and large rocks are the first to lose their snow. The ground snowpack is also shallower due to the warming effects of the sun. In the southern hemisphere, all of these effects will be the opposite.

NAVIGATION METHODS

7-47. There are three primary means of navigation generally available to the isolated person: global positioning system (GPS), dead reckoning, and terrain association.

GLOBAL POSITIONING SYSTEM

7-48. The GPS is a space-based, global, all-weather, continuously available, radio positioning navigation system. It is highly accurate in determining position location derived from signal triangulation from a satellite constellation system. It is capable of determining grid and geographic coordinates (i.e. latitude and longitude) as well as altitude of the individual user. It is fielded in handheld, man pack, vehicular, aircraft, and watercraft configurations.

7-49. The GPS receives and processes data from a "constellation" of satellites on a simultaneous or sequential basis. It transmits a radio signal to each satellite which, by measuring the time taken for the signal to reach it, is able to determine the distance between the signal's source and its receiver. By comparing the distances between multiple satellites (in conjunction with information about the position of the satellites themselves) the system is able to determine where these distance radii intersect and thus the precise position of the signal's source. The system then transmits this geolocation information to the source device as geographic or military grid coordinates. There are several considerations for the employment of GPS by the isolated person:

- GPS has an electronic compass and the ability to store waypoints and enter routes of movement to follow. As an electronic device, it is vulnerable to enemy jamming, spoofing and interference.
- Protect the GPS and conserve battery life to increase its operational life during isolation.
- Zero out (remove all user data from, or reset) the GPS prior to possible capture to mitigate its exploitation value.
- Use GPS in conjunction with other navigation tools to minimize the chances of becoming lost or unable to correctly execute ISG/EPA.

DEAD RECKONING

7-50. Dead reckoning consists of two fundamental steps. The first step is to use a protractor and graphic scales to determine the direction and distance from one point to another on a map. The second step is to use a compass and some means of measuring distance to apply this information on the ground. In other words, it begins with the determination of a polar coordinate on a map and ends with the act of finding it on the ground.

7-51. Dead reckoning along a given route is the application of the same process used by a mapmaker to establish a measured line of reference upon which to construct the framework of the map. Resection[1] or intersection[2] can be easily undertaken at any time to either determine or confirm precise locations along or near your route. Between these position fixes, isolated persons can establish their location by measuring or estimating the distance traveled along the azimuth being followed from the previous known point. They might use pacing, a vehicle odometer, or the application of elapsed time for this purpose, depending upon the situation.

7-52. Most dead reckoned movements do not consist of single straight-line distances because the tactical and navigational aspects of the terrain, the enemy situation, natural and man-made obstacles, time, and safety factors cannot be ignored. Another reason most dead reckoning movements are not single straight-line distances is because compasses and pace counts are imprecise measures. Error from them compounds over distance; therefore, isolated persons could soon be far from their intended route even if they performed the procedures correctly. The only way to counteract this phenomenon is to reconfirm the location by terrain association (see page 240) or resection. Routes planned for dead reckoning generally consist of a series of straight-line distances between several checkpoints.

7-53. There are two advantages to dead reckoning. First, dead reckoning is easy to teach and learn. Second, it can be a highly accurate way of moving from one point to another if done carefully over short distances, even where few external cues are present to guide movements.

7-54. During daylight, across open country, along a specified magnetic azimuth, isolated persons should never walk with the compass in the open position and in front of them. Because the compass will not stay steady or level, it does not give an accurate reading when held or used this way. Begin at the start point and face with the compass in the proper direction, then sight in on a landmark located on the correct azimuth to be followed. Close the compass and

---

1. Resection is a method of determining an unknown position (your location) by reference to two or more known positions. To determine your position by resection, choose two visible, mapped objects. Take a bearing from your position to the first object, then draw a line on your map from the object on that bearing. Repeat this for the second object. Your location is the place on your map where the two (or more) bearing lines intersect.

2. Intersection is similar to resection, except that you determine the position of an unknown, unmapped distant object by yourself occupying two or more known positions in sequence. To determine a distant object's position by intersection, choose two or more known, mapped points on your route. When you reach the first point, take a bearing from your position to the unmapped object, then draw a line on your map from your position on that bearing. Repeat this when you reach the second mapped location. The distant, unmapped position is the place on your map where the two (or more) bearing lines intersect.

proceed to that landmark. Repeat the process as many times as necessary to complete the straight-line segment of the route.

7-55. The landmarks selected for these purposes are called "steering marks", and their selection is crucial to success in dead reckoning. Steering marks should never be determined from a map study. They are selected as the march progresses and are commonly on or near the highest points visible along the azimuth line being followed. They may be uniquely shaped trees, rocks, hilltops, posts, towers, and buildings—anything that can be easily identified. If isolated persons do not see a good steering mark to the front, they might use a back azimuth to some feature behind them until a good steering mark appears out in front. Characteristics of a good steering mark include the following:

- It must have some characteristics, such as color, shade of color, size, or shape (preferably all four), that assures you that it will continue to be recognized as you approach it.

- It is the most distant object available along the line of the march. This enables you to move farther with fewer references to the compass. If there are many options, you should select the highest object. A higher mark is not as easily lost to sight as is a lower mark that may blend into the background as it is approached. A steering mark should be continuously visible as you move toward it.

- A steering mark selected at night must have even more unique shapes than one selected during daylight. As darkness approaches, colors disappear and objects appear as black or gray silhouettes. Instead of seeing shapes, only the general outlines are visible; they may appear to change as you move and see the object from slightly different angles.

7-56. Dead reckoning without natural steering marks is used when the travel area is devoid of features, or when visibility is poor. At night, it may be necessary to send a member of the unit out in front of the unit's position to serve as a steering mark in order to proceed. The position should be as far out as possible to reduce the number of chances for error during movement. Arm-and-hand signals or a radio may be used in placing the isolated person on the correct azimuth. After being properly located, move forward to this position and repeat the process until some natural steering marks can be identified (or the objective is reached). This process has the advantage of reducing error due to the fact that your personnel occupy both the known-good position and the steering mark position simultaneously: as you reach the steering mark and repeat the process, you can therefore have a reasonably high degree of confidence in your route's accuracy.

7-57. When handling obstacles and detours on the route, follow these guidelines:

- When an obstacle forces you to leave your original line of march and take up a parallel one, always return to the original line as soon as the terrain or situation permits.

- To turn clockwise (right) 90°; 90° must be added to your original azimuth. To turn counterclockwise (left) 90° from your current direction, 90° must be subtracted from your present azimuth.

- When making a detour, be certain that only paces taken toward the final destination are counted as part of the forward progress. The forward progress paces should not be confused with the local pacing that takes place perpendicular to the route in order to avoid the problem area and in returning to the original line of march after the obstacle has been passed.

7-58. You can use the deliberate offset technique. Highly accurate distance estimates and precision compass work may not be required for a deliberate offset technique if the destination or an intermediate checkpoint is located on or near a large linear feature that runs nearly perpendicular to the direction of travel. Examples include roads or highways, railroads, power transmission lines, ridges, or streams (collectively known as "handrails"—see page 241). In these cases, you should apply a deliberate error (offset) of about 10° to the azimuth you planned to follow and then move, using the lensatic compass as a guide, in that direction until you encounter the linear feature. You will know exactly which way to turn (left or right) to find your destination or checkpoint, depending upon which way you planned your deliberate offset.

7-59. Because no one can move along a given azimuth with absolute precision, it is better to plan a few extra steps than to begin an aimless search for the objective once reaching the linear feature. This method also copes with minor compass errors and the slight variations that always occur in the earth's magnetic field.

7-60. There are disadvantages to dead reckoning. The farther the movement by dead reckoning without confirming the position in relation to the terrain and other features, the more errors that will accumulate in the movements. Therefore, you should confirm and correct your estimated position whenever you encounter a known feature on the ground that is also on the map. Periodically, complete a resection using two or more known points to pinpoint and correct your position on the map. Pace counts or any type of distance measurement should begin anew each time your position is confirmed on the map.

TERRAIN ASSOCIATION

7-61. The technique of moving by terrain association is more forgiving of mistakes and far less time-consuming than dead reckoning. It best suits those situations that call for movement from one area to another. Once an error has been made in dead reckoning, the isolated person is off the track. However, errors made using terrain association are easily corrected, because isolated persons are comparing what they expected to see from the map to what they do see on the ground.

7-62. Errors are anticipated and will not go unchecked. Isolated persons can easily make adjustments based upon what they encounter. Periodic position fixing through either plotted or estimated resection will also make it possible to correct movements.

Identifying and Locating Selected Features

7-63. Being able to identify and locate selected features, both on the map and on the ground, are essential to the success in moving by terrain association. The following rules may prove helpful:

- Be certain the map is properly oriented when moving along the route and use the terrain and other features as guides. The orientation of the map must match the terrain or it can cause confusion.
- When locating and identifying features to be used to guide movement, look for the steepness and shape of slopes, the relative elevations of various features, and the directional orientations in relation to the isolated person's position and to the position of other features that can be seen.
- Make use of the additional cues provided by hydrography and vegetation.

Using Handrails, Catching Features, and Navigational Attack Points

7-64. The following paragraphs discuss how to use handrails, catching features, and attack points to determine direction.

Handrails

7-65. "Handrails" are linear features, such as roads or highways, railroads, power transmission lines, ridgelines, or streams that run roughly parallel to your direction of movement. Instead of using precision compass work, isolated persons can "rough compass" (use the linear feature to follow the general compass direction) without the use of steering marks for as long as the feature travels with them on their right or left. It acts as a handrail to guide the way.

Catching Features

7-66. When you reach the point where either your route or the handrail changes direction, you must be aware that it is time to go your separate ways. Some prominent feature located near this point is selected to provide this warning. This is called a "catching feature"; it can also be used to tell when you have gone too far, acting as a backstop.

Navigational Attack Points

7-67. The catching feature may also be your "navigational attack point"; this point is the place where area navigation ends and point navigation begins. From this last easily-identified checkpoint, move cautiously and precisely along a given azimuth for a specified distance to locate the final objective. The selection of this navigational attack point is important. A distance of about 500 meters or less from the final objective is most desirable.

Combining Techniques

7-68. The most successful navigation is obtained by combining the techniques described above. Constant orientation of the map and continuous observation of the terrain in conjunction with compass-read azimuths, and distance traveled on the ground compared with map distance, when used together, make reaching a destination more certain. Isolated persons should not depend entirely on compass navigation or map navigation.

MAP ORIENTATION

7-69. Prior to navigation, you will orient the map. A map is oriented when it is in a horizontal position with its north and south corresponding to the north and south on the ground.

Using a Compass

Note: Maps are drawn to be a specific representation of a piece of the earth's surface. This representation is oriented (or pointed) to true north: i.e., in the direction of the geographic North Pole, which is located along the earth's rotational axis. In contrast, the compass points to magnetic north (or the north magnetic pole). This is in the general direction of true north, but depending on where the map is drawn to on the earth's surface there is a variation between the two. This is an easterly or westerly variation and will be represented in degrees. The declination diagram in the map's information section will provide the angular relationship between true north and magnetic north. This allows for making an adjustment to the map that is known as orienting the map. The best way to orient a map is with a compass. For use of a lensatic compass, see TC 3-25.26[1].

7-70. When orienting a map with a compass, remember that the compass measures magnetic azimuths. Since the magnetic arrow points to magnetic north, pay special attention to the declination diagram. Two techniques to orient the map are used—

- **First technique**:

1. Also available from Carlile Media: ISBN 1977649270.

- Look at the map's declination diagram. Determine the direction of the declination and its value (usually marked "G-M angle").
- Place the map in a horizontal position.
- Take the straightedge on the left side of the compass and place it alongside the north-south grid line (with the cover of the compass pointing toward the top of the map). This procedure places the fixed black index line of the compass parallel to north-south grid lines of the map.
- Keeping the compass aligned as directed above, rotate the map and compass together, until the magnetic arrow of the compass is below its fixed black index line.
- The map and compass are correctly oriented when they match the G-M angle given by the map. To check this when the magnetic north arrow on the map is to the left of the grid north, check the compass reading to see if it equals the G-M angle given in the declination diagram. To check this when the magnetic north arrow is to the right of the grid north, check the compass reading to see if it equals 360 degrees minus the G-M angle.

- **Second technique**:
  - Look at the map's declination diagram. Determine the direction of the declination and its value (G-M angle).
  - Place the map in a horizontal position.
  - Using any north-south grid line on the map as a base, draw a magnetic azimuth equal to the G-M angle given in the declination diagram, using your protractor.
  - If the declination is easterly (right), the drawn line is equal to the value of the G-M angle: align the straightedge on the left side of the compass alongside the drawn line on the map. Rotate the map and compass until the magnetic arrow of the compass is below the fixed black index line: the map is now oriented.
  - If the declination is westerly (left), the drawn line will equal 360 degrees minus the value of the G-M angle: align the straightedge on the left side of the compass alongside the drawn line on the map. Rotate the map and compass until the magnetic arrow of the compass is below the fixed black index line: the map is now oriented.

Note: Use caution to ensure nothing (metal, mine and ore) in the area will alter the compass reading.

7-71. The lensatic compass has a needle with a north direction marked on the bottom inside of the compass. A button or wrist compasses may have floating dials or floating needles. Orienting a map with a floating needle compass is similar to the method used with the floating dial.

Figure 7-13. Map Orientation Examples

A

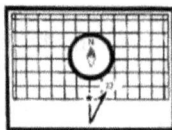

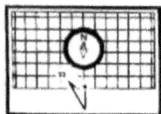

Floating needle compass and map aligned to magnetic North

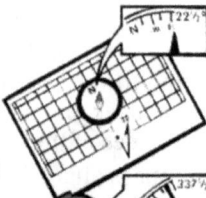

Map is oriented to 22 1/2° Easterly magnetic variation with floating needle compass

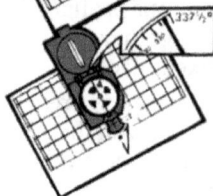

Map is oriented to 22 1/2° Easterly magnetic variation with floating dial compass

B

Floating needle compass and map aligned to magnetic North

Map is oriented to 22 1/2° Westerly magnetic variation with floating needle compass

Map is oriented to 22 1/2° Westerly magnetic variation with floating dial compass

---

Without a Compass

7-72. When a compass is not available, map orientation requires a careful examination of the map and the ground to find linear features common to both, such as roads, railroads, and power lines. The map is oriented by aligning a feature on the map with the same feature on the ground. Orientation by this method must be checked to prevent the reversal of directions, which may occur if only one linear feature is used. This reversal may be prevented by aligning two or more map features (terrain or man-made). If no second linear feature is visible but the map user's position is known, a prominent object may be used. With the prominent object and the user's position connected with a straight line on the map, the map is rotated until the line points toward the feature.

7-73. If two prominent objects are visible and plotted on the map and the user's position is not known, move to one of the plotted and known positions, place the straightedge or protractor on the line between the plotted positions, and turn the protractor and the map until the other plotted and visible point is seen along the edge. The map is then oriented (see Figure 7-14).

Figure 7-14. Orienting a Map to Terrain Features

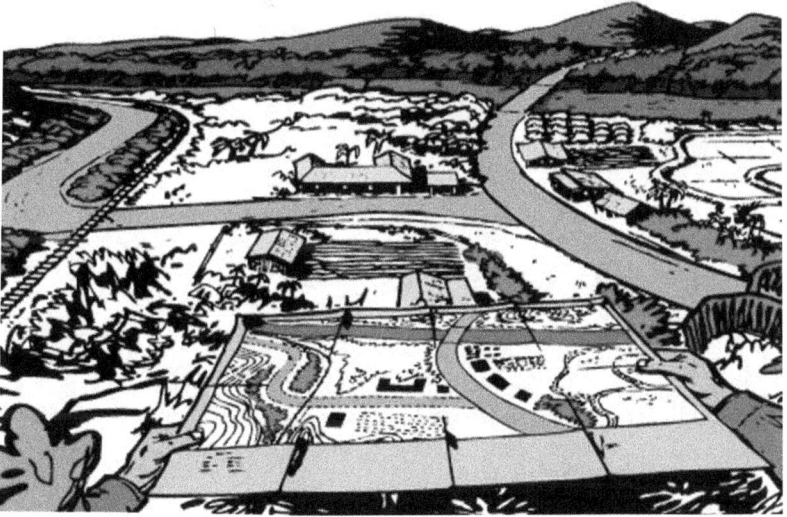

TERRAIN FEATURES

7-74. Major terrain features include—

* **Hills.** A hill is an area of high ground. From a hilltop, the ground slopes down in all directions. A hill is shown on a map by contour lines forming concentric circles. The inside of the smallest closed circle is the hilltop (see Figure 7-15 on page 246).

Figure 7-15. A Hill

Hill

- **Saddles**. A saddle is a dip or low point between two areas of higher ground. A saddle is not necessarily the lower ground between two hilltops; it may be simply a dip or break along a level ridge crest. In a saddle, there is high ground in two opposite directions and lower ground in the other two directions. It is normally represented as an hourglass (see Figure 7-16).

Figure 7-16. A Saddle

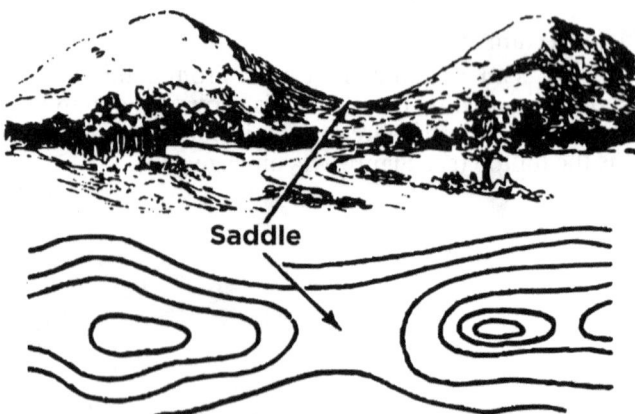

Saddle

- **Valleys**. A valley is a stretched-out groove in the land, usually formed by streams or rivers. A valley begins with high ground on three sides and usually has a course of running water through it. In a valley, three directions offer high ground, while the fourth direction offers low ground. Depending on its size and where a person is standing, it may not be obvious that there is high ground in the third direction, but water flows from higher to lower ground. Contour lines forming a valley are either U-shaped or V-shaped. To determine the direction water is flowing, look at the contour lines. The closed end of the contour line (U or V) always points upstream or toward high ground (see Figure 7-17).

Figure 7-17. A Valley

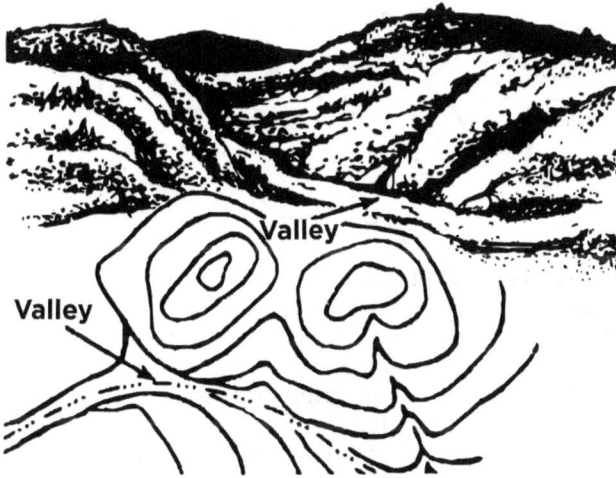

- **Ridges**. A ridge is a sloping line of high ground. On the centerline of a ridge, there will normally be low ground in three directions and high ground in one direction with varying degrees of slope. If you cross a ridge at right angles, you will climb steeply to the crest and then descend steeply to the base. When you move along the path of the ridge, depending on the geographic location, there may be either an almost unnoticeable slope or a very obvious incline. Contour lines forming a ridge tend to be U-shaped or V-shaped. The closed end of the contour line points away from high ground (see Figure 7-18 on page 248).

Figure 7-18. A Ridge

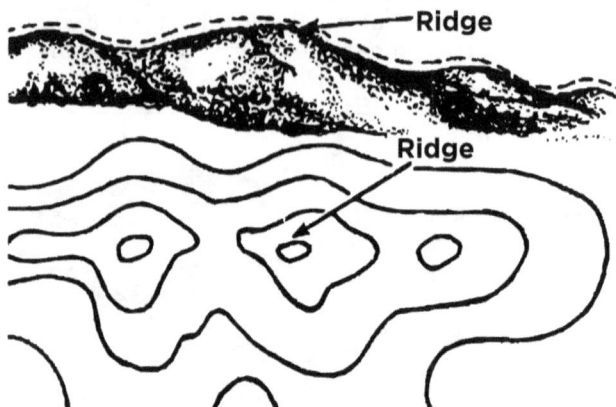

• **Depressions**. A depression is a low point in the ground or a sinkhole. It could be described as an area of low ground surrounded by higher ground in all directions, or simply as a hole in the ground. Usually only depressions whose depth is equal to or greater than the contour interval will be shown. On maps, depressions are represented by closed contour lines that have tick marks pointing toward low ground (see Figure 7-19).

Figure 7-19. A Depression

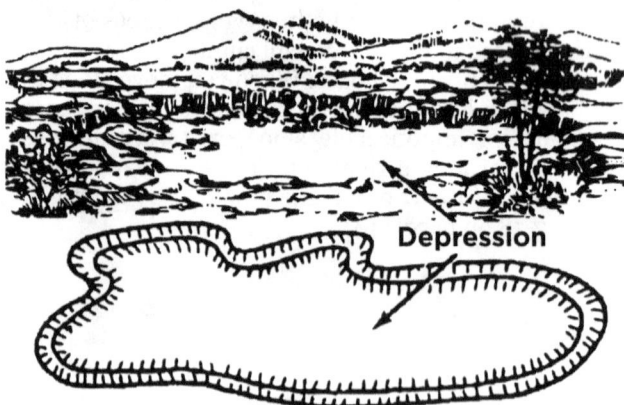

7-75. Minor terrain features include—

• **Draws**. A draw is a stream course that is less developed than a valley. In a draw, there is essentially no level ground; therefore, there is little or no maneuver room within its confines. In a draw, the ground slopes upward in three directions and downward in the other direction. A draw could be considered as the initial formation of a valley. The contour lines depicting a draw are U-shaped or V-shaped, pointing toward high ground (see Figure 7-20).

Figure 7-20. A Draw

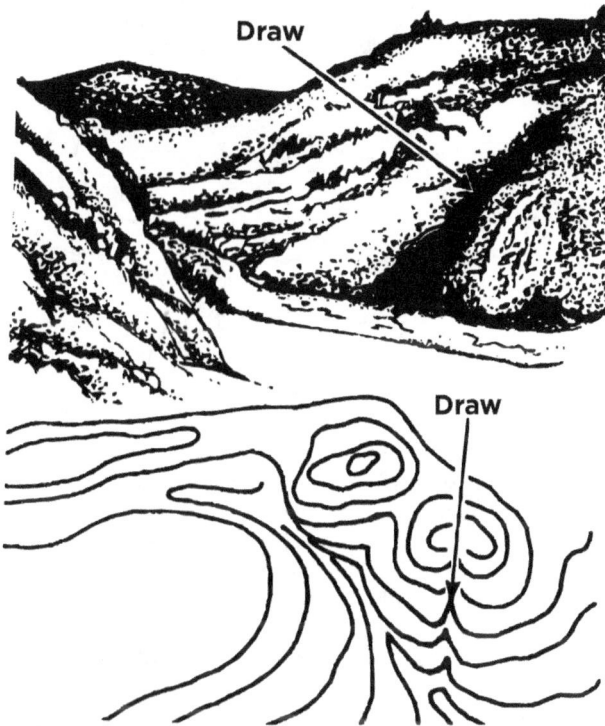

- **Spurs**. A spur is a short, continuously sloping line of higher ground, normally jutting out from the side of a ridge. A spur is often formed by two roughly parallel streams cutting draws down the side of a ridge. The ground will slope down in three directions and up in one. Contour lines on a map depict a spur with the U or V pointing away from high ground (see Figure 7-21 on page 250).

Figure 7-21. A Spur

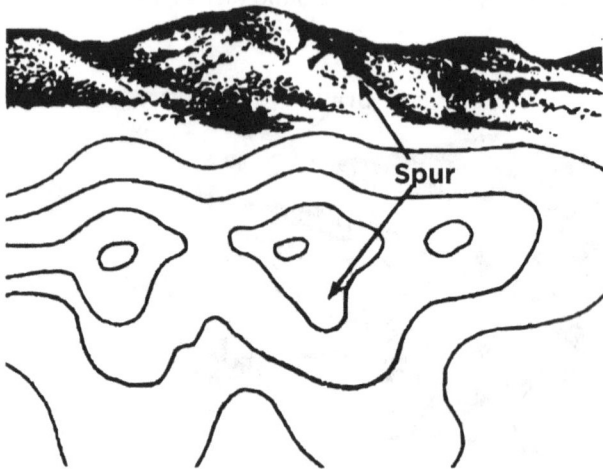

- **Cliffs.** A cliff is a vertical or near-vertical feature; it is an abrupt change in the height of the land. When a slope is so steep that the contour lines converge into one "carrying" contour of contours, this last contour line has tick marks pointing toward low ground). Cliffs are also shown by contour lines very close together and, in some instances, touching each other (see Figure 7-22).

Figure 7-22. A Cliff

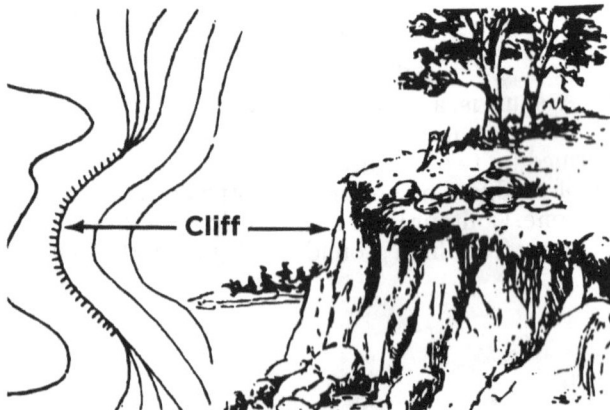

7-76. Supplementary terrain features include—
- **Cuts.** A cut is a man-made feature resulting from cutting through raised ground, usually to form a level bed for a road or railroad track. Cuts are shown on a map when they are at least 10 feet high, and they are drawn with a contour line along the cut line. This contour line extends the length of the cut and has tick marks that

extend from the cut line to the roadbed, if the map scale permits this level of detail.

- **Fills**. A fill is a man-made feature resulting from filling a low area, usually to form a level bed for a road or railroad track. Fills are shown on a map when they are at least 10 feet high, and they are drawn with a contour line along the fill line. This contour line extends the length of the filled area and has tick marks that point toward lower ground. If the map scale permits, the length of the fill tick marks are drawn to scale and extend from the base line of the fill symbol.

MEASURING DISTANCES

7-77. There are different ways of measuring distances. Several techniques are included in this section.

GRAPHIC SCALES

7-78. On most maps, there is another method of determining ground distance. It is by means of the graphic (bar) scales. A graphic scale is a ruler printed on the map on which distances on the map may be measured as actual ground distances. To the right of the zero (0), the scale is marked in full units of measure and is called the "primary scale." The part to the left of zero (0) is divided into tenths of a unit and is called the "extension scale." Most maps have three or more graphic scales, each of which measures distance in a different unit of measure (see Figure 7-23).

Figure 7-23. Graphic Scale

EXTENSION SCALE   **PRIMARY SCALE**

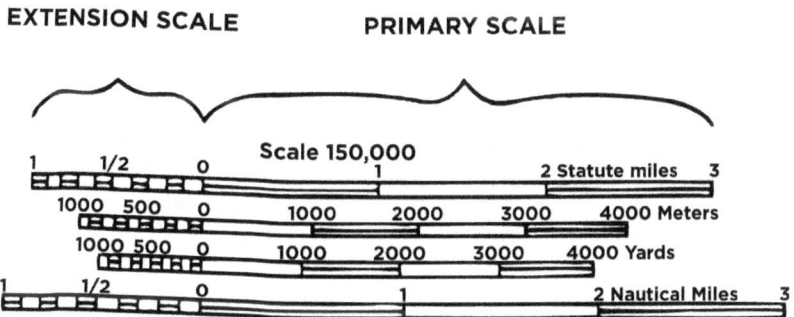

Ground Distance—Straight Line

7-79. To determine a straight-line ground distance between two points on a map, lay a straight-edged piece of paper on the map so that the edge of the paper touches both points. Mark the straight edge of the paper at each point. Move the paper down to the graphic scale and read the ground distance between the points. Use the scale marked in the unit of measure desired (see Figure 7-24 on page 252).

Figure 7-24. Measure Map Distance—Straight Line

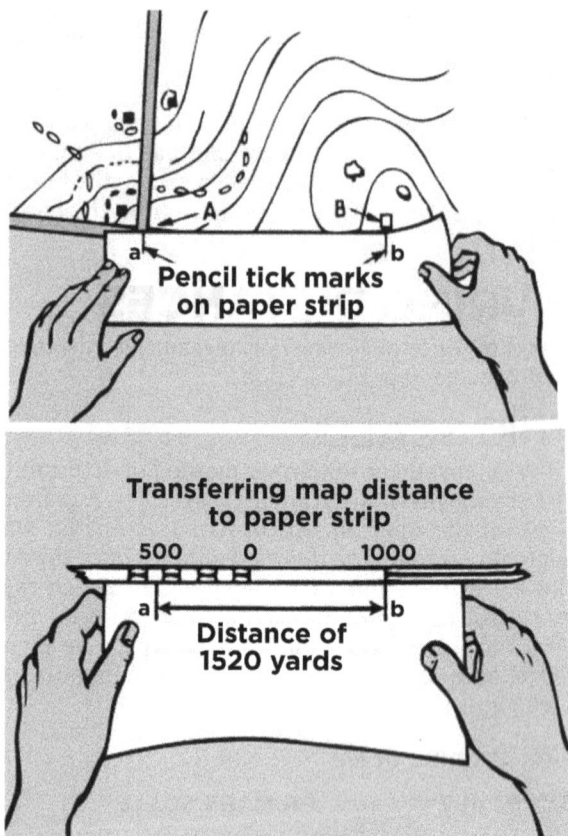

Pencil tick marks
on paper strip

Transferring map distance
to paper strip

500   0   1000

a |← →| b

Distance of
1520 yards

Ground Distance—Curved Line

7-80. To measure distance along a winding road, stream, or any other curved line, the straight edge of a piece of paper is used. Mark one end of the paper and place it at the point from which the curved line is to be measured. Align the edge of the paper along a straight portion and mark both the map and the paper at the end of the aligned portion. Keeping both marks together, place the point of the pencil on the mark on the paper to hold it in place. Pivot the paper until another approximately straight portion is aligned and again mark on the map and the paper. Continue in this manner until measurement is complete. Then place the paper on the graphic scale and read the ground distance (see Figure 7-25 on page 253).

Figure 7-25. Measure Map Distance—Curved Line

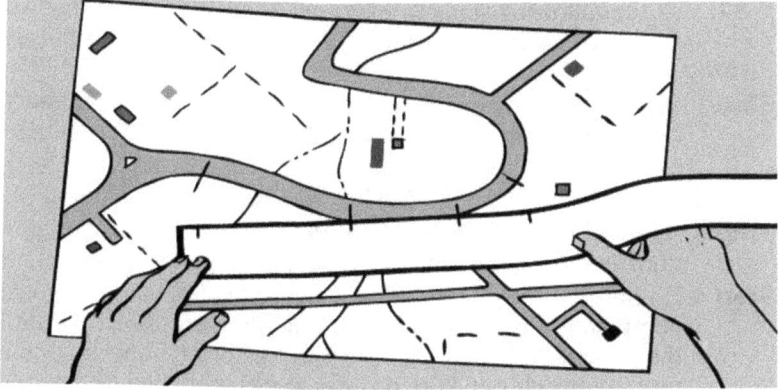

MEASURING AZIMUTH

7-81. Use the following steps to determine the grid azimuth of a line from one point to another on a map, for example, from point A to point B, or from point C to point D (see Figure 7-26):

Step 1 Draw a line connecting the two points.

Step 2 Place the index of the protractor at the point where the line crosses a vertical (north-south) grid line.

Step 3 Keeping the index at this point, align the 0° to 180° line of the protractor on the vertical grid line.

Step 4 Read the value of the angle from the scale; this is the grid azimuth to the point.

Figure 7-26. Measuring Azimuth on a Map

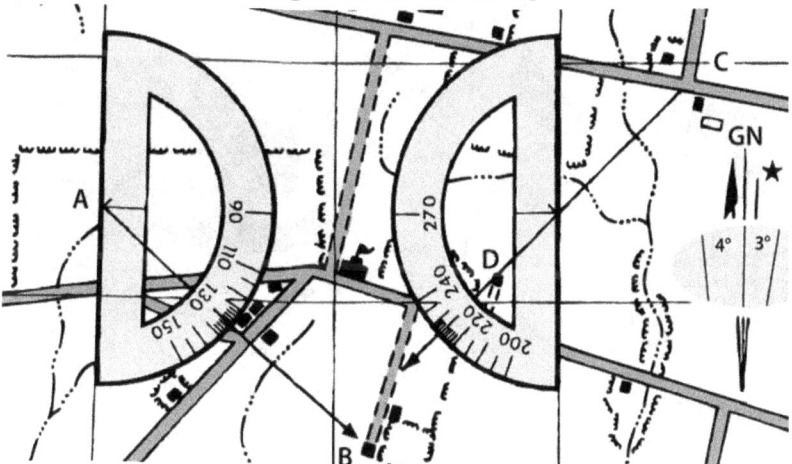

PLOTTING A DIRECT LINE

7-82. Use the following steps to plot a direction line from a known point on a map, using a provided direction or grid azimuth (see Figure 7-27):

Step 1 Convert the provided direction to a grid azimuth, if necessary (for example, if you are provided with the direction northeast, the azimuth would be 45°).

Step 2 Imagine a north-south grid line through the known point.

Step 3 Approximately align the 0° to 180° line of the protractor in a north-south direction through the known point.

Step 4 Holding the 0° to 180° line of the protractor on the known point, slide the protractor in the north-south direction until the horizontal line of the protractor (connecting the protractor index and the 90° tick mark) is aligned on an east-west grid line and draw a vertical line connecting 0°, the known point, and 180°. This vertical line allows you to accurately move the protractor vertically in the following step.

Step 5 Holding the 0° to 180° line on the north-south line, slide the protractor index to the known point.

Step 6 Make a mark on the map at the required angle.

Step 7 Draw a line from the known point through the mark made on the map.

Figure 7-27. Plotting a Direction

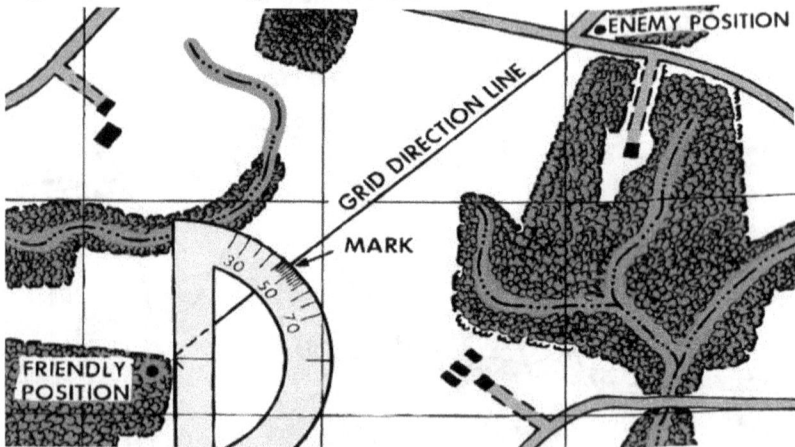

MEASURING DISTANCES BY PACES

7-83. A pace is equal to the distance covered every time the same foot touches the ground. To measure distance, count the number of paces, estimate the distance traveled, and apply that to the map and route of movement. Distances measured this way are only approximate. Before using dead reckoning navigation, it is important for each person to establish the length of his or her average pace. This is done by pacing a measured course many times and computing the mean. In the field, an average pace must often be adjusted because of the following conditions—

- **Slopes**. The pace lengthens on a downgrade and shortens on an upgrade.
- **Winds**. A headwind shortens the pace while a tailwind increases it.
- **Surfaces**. Sand, gravel, mud, and similar surface materials tend to shorten the pace.
- **Elements**. Snow, rain, or ice reduces the length of the pace.
- **Clothing**. Excess weight of clothing shortens the pace while the type of shoes affects traction and therefore pace length.

FIELD-EXPEDIENT GUIDELINES FOR DISTANCE TO A LANDMARK

7-84. Isolated persons must be able to estimate distance between themselves and an object. One way this can be done is by looking at trees. If you can see individual tree branches, you can estimate that you are about 0~0.6 mile away from that landmark. When only individual trees are visible, you are about 1~3 kilometers away. If you can see the trees as a sort of shag-carpet look, the landmark would be about 3~5 kilometers away. A smooth looking carpet appearance indicates that the landmark is about 5~7 kilometers away. A bluish tint or haze in the distance indicates the landmark is about 7~10 kilometers away.

7-85. Darkness presents its own challenges for land navigation because of limited or no visibility. However, the techniques and principles are the same as those used for day navigation. Success in nighttime land navigation depends on rehearsals during the planning phase before the intended movement, such as detailed analysis of the map to determine the type of terrain in which the navigation is going to take place, and the predetermination of azimuths and distances. Night vision devices can greatly enhance night navigation.

7-86. Night movements are considerably slower than day movements, but METT-TC[1] considerations may require movement during darkness. Pace counts increase during darkness (as your pace length will decrease due to low visibility), map checks will take longer, accountability is more difficult. The use of flashlights is not recommended. If an artificial light source is used, ensure it is pointed straight down; during a map check it is advisable to cover yourself

---

1. Mission, enemy, terrain, troops available, time, and civilian considerations.

with a poncho or other material. Noise tends to carry at night and movement through a highly vegetated area may produce more noise than during daylight. In this case it is advisable to alter the route to stay well away from any danger areas or cease movement altogether.

7-87. Navigation using the stars is recommended in some areas. The four cardinal directions can also be obtained at night using the same technique described for the shadow-tip method—just use the moon instead of the sun (in this case, the moon must be bright enough to cast a shadow).

CHAPTER 8

SURVIVAL EQUIPMENT

This chapter discusses the employment and maintenance of survival equipment and construction of field-expedient weapons. Weapons serve a dual purpose: they are used both to collect and prepare food and also to provide for self-defense.

MAINTAINING EQUIPMENT

8-1. Temperature extremes, both hot or cold, are stressful on equipment and can impede its use. For example, electronics require batteries, which have a limited amount of stored energy. Radios, GPSs, flashlights, cell phones, beacons, and blue-force trackers all need battery power to function. If exposed to temperature extremes, a battery's power can be degraded and internal parts can become damaged. This could leave isolated persons stranded during critical times when the equipment in needed. Extreme heat can cause batteries to overheat and potentially explode, or degrade the stored-up energy. Batteries should be kept out of direct sunlight and away from the open flame of any fire.

PROTECTION

8-2. In extreme cold weather, electronic equipment and spare batteries should be kept between layers of clothing to stay warm or in shelters out of the cold. Moisture has damaging effects on all types of equipment, and isolated persons should do everything possible to keep their items dry. Humidity encourages mold in enclosed spaces such as optics. Items such as optics should be kept in as dry a condition as possible so condensation does not form inside the sealed compartments. If equipment becomes wet, it should be opened up and dried out thoroughly. Metal items can rust and moving parts will need lubrication if equipment is not completely dried. Metal items should be cleaned and lightly lubricated when not required for immediate use.

LUBRICANTS AND GLUE

8-3. Animal fat can be used as a lubricant and/or as fuel for an improvised lamp or heater. A strong glue can be made that is perfect for gluing organic material together. It is made by boiling animal hide scrapings or rawhide until it dissolves into a thin liquid. This glue must be used while it is hot, as it will set quickly upon cooling. This type of glue will stretch if it gets wet and it will break down if it is exposed to high heat.

8-4. Plants that produce sap, mainly conifers like pine and spruce trees, can produce a glue that is waterproof. The sap from these trees is found as blisters of tree resin on the trunk and will also collect near wounds in the bark of the tree. These resin accumulations can be collected and placed in a container and cooked-down over a fire. Once the material starts to melt you will notice a turpentine smell. Do not allow the resin to boil because this will weaken the product. The resin should be covered as it is cooked down because it is highly flammable.

8-5. Once the liquid glue is cooked down, it should be filtered to remove large impurities. You should add some type of tempering material. Tempering material makes the glue stronger and more stable. Types of tempering material are charcoal from the fire, beeswax, and animal scat (droppings) from herbivores like rabbits or deer. For the right amount of tempering material to add, mix in a small amount, test the glue if it is sticky and soft, then add more material till it thickens to the desired consistency. The final product should be separated in usable portions and stored until needed for a project.

8-6. To use the glue, heat the glue and apply it to the surface of the project with a stick. Heating the project's surface before gluing or repairing the joint of the project increases the effectiveness and strength of the glue joint.

FIELD-EXPEDIENT WEAPONS

8-7. In survival situations, you may have to fashion any number and type of field-expedient tools and equipment to survive

IMPROVISING

8-8. The need for an item must outweigh the work involved in making it. Follow these simple principles when improvising:

- Is it necessary or just nice to have?
- Determine the actual need.
- Inventory possessions and available natural and man-made resources.

- Consider all alternatives.
- Select the alternative providing the most efficient use of materials, time, and energy.
- Plan all construction to ensure that it is durable and safe.

STAFFS

8-9. A staff should be one of the first tools obtained. For walking, it provides support and helps in ascending and descending steep slopes. It also provides some weapons capabilities if used properly, especially against snakes and dogs. A staff should be about the same height as the person using it or at least eyebrow height. It should be no larger than you can effectively wield when tired and undernourished. A staff provides invaluable eye protection when moving through heavy brush and thorns in darkness. It can also be used for self-protection by thrusting, striking, and blocking potential threats.

CLUBS

8-10. Clubs should be held, but never be thrown. However, a club can extend your area of defensive reach beyond your fingertips. It also serves to increase the force of a blow without injuring the wielder. The three basic types of clubs are explained as follows—

Simple Club

8-11. A simple club is a staff or branch. It must be short enough to swing easily, but long enough and strong enough to damage whatever it hits. Its diameter should fit comfortably in the palm, but it should not be so thin as to allow the club to break easily upon impact. A straight-grained hardwood is best, if available. A club is lightweight and effective as an impact weapon and can be swung at different angles to strike and block potential threats.

Weighted Club

8-12. A weighted club is any simple club with a weight on one end. The weight may be a natural weight, such as a knot on the wood, or something added, such as a stone lashed to the club. To make a weighted club, first find a stone that has a shape that will allow easy and secure lashing to the club. A stone with a slight hourglass shape works well, as lashing will not easily slip off. If a suitably shaped stone cannot be found, fashion a groove or channel into the stone by "pecking," or repeatedly rapping the club stone with a smaller hard stone.

8-13. Next, find a piece of wood that is the right length. Again, use straight-grained hardwood is if possible. The length of the wood should feel comfortable in relation to the weight of the stone, so the club can be swung with speed and accuracy. Finally, lash the stone to the handle using a technique shown in Figure 8-1. The technique used will depend on the type of handle chosen.

Figure 8-1. Weighted Clubs

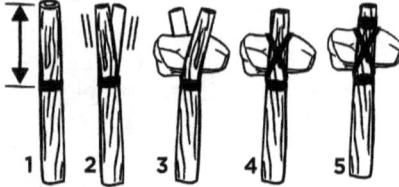

15~20 cm
(6~8 in)

1   2   3   4   5

Split-handle technique

1 Wrap lashing
2 Split end of lashing
3 Insert stone
4 Lash securely above, below and across stone
5 Bind split end tightly to secure stone

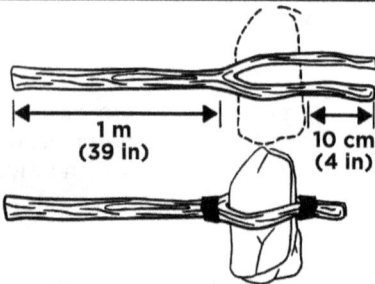

1 m
(39 in)

10 cm
(4 in)

Starting at crotch lash securely to prevent splitting

Forked-branch technique

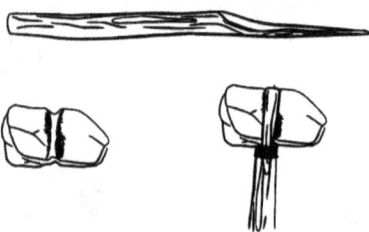

Wrapped-handle technique

1 Take hardwood 1 m (39 in) long and 2.5 cm (1 in) in diameter and shave end to about half the diameter
2 Take about a 1.8 kg (4 pounds) stone with "pecked groove" and wrap the shaved end around the stone
3 Lash securely

Sling Club

8-14. A sling club is another type of weighted club. A weight hangs 3 to 4 inches from the handle by a strong, flexible lashing. This type of club both extends the user's reach and multiplies the force of the blow (see Figure 8-2 on page 261).

Figure 8-2. Sling Club

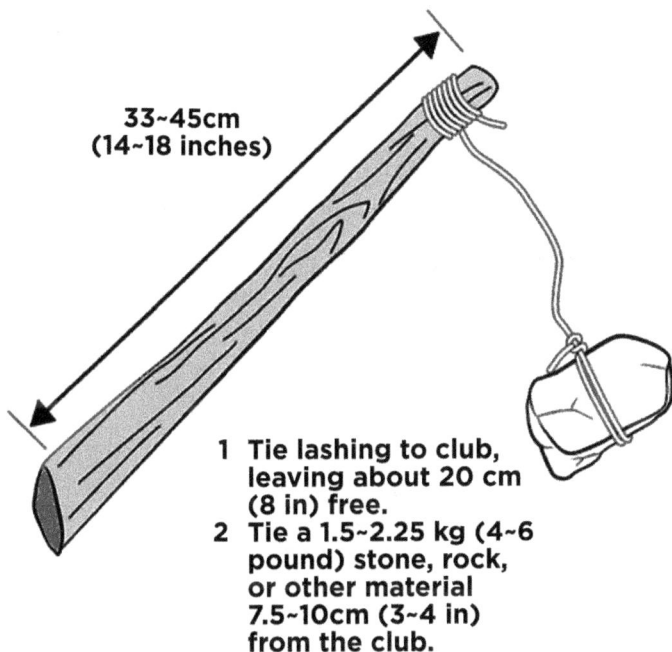

**33~45cm
(14~18 inches)**

1 **Tie lashing to club,
leaving about 20 cm
(8 in) free.**
2 **Tie a 1.5~2.25 kg (4~6
pound) stone, rock,
or other material
7.5~10cm (3~4 in)
from the club.**

EDGED WEAPONS

8-15. Knives, spear blades, and arrow points fall under the category of edged weapons. The following Paragraphs explain how to make such weapons.

Knives

8-16. A knife has three basic functions. It can puncture, slash or chop, and cut. A knife is also an invaluable tool used to construct other survival items. In a survival situation, you may find yourself without a knife or you may need another type of knife or a spear. To improvise, you can use stone, bone, wood, or metal to make a knife or spear blade.

Stone

8-17. A stone blade will make a good puncturing tool and a good chopping tool but in general will not hold a fine edge (see Figure 8-3 on page 262). However, some stones such as obsidian, chert, or flint can have very fine edges. To make a stone knife, a sharp-edged piece of stone, a chipping tool, and a flaking tool will be needed. A chipping tool is a light, blunt-edged tool used to break off small pieces of stone. A flaking tool is a pointed tool used to break off thin, flattened pieces of stone. A chipping tool can be made from wood, bone, or metal and a flaking tool can be made from bone, antler tines, or soft iron.

8-18. To make a stone blade, begin by roughing out the desired shape on the sharp piece of stone, using the chipping tool. Try to make the knife thin. Press the flaking tool against the edges. This action will cause flakes to come off the opposite side of the edge, leaving a razor-sharp edge. Use the flaking tool along the entire length of the edge to be sharpened. Eventually, this will produce a very sharp cutting edge that can be used as a knife. Lash the blade to some type of hilt.

Figure 8-3. Stone Knife

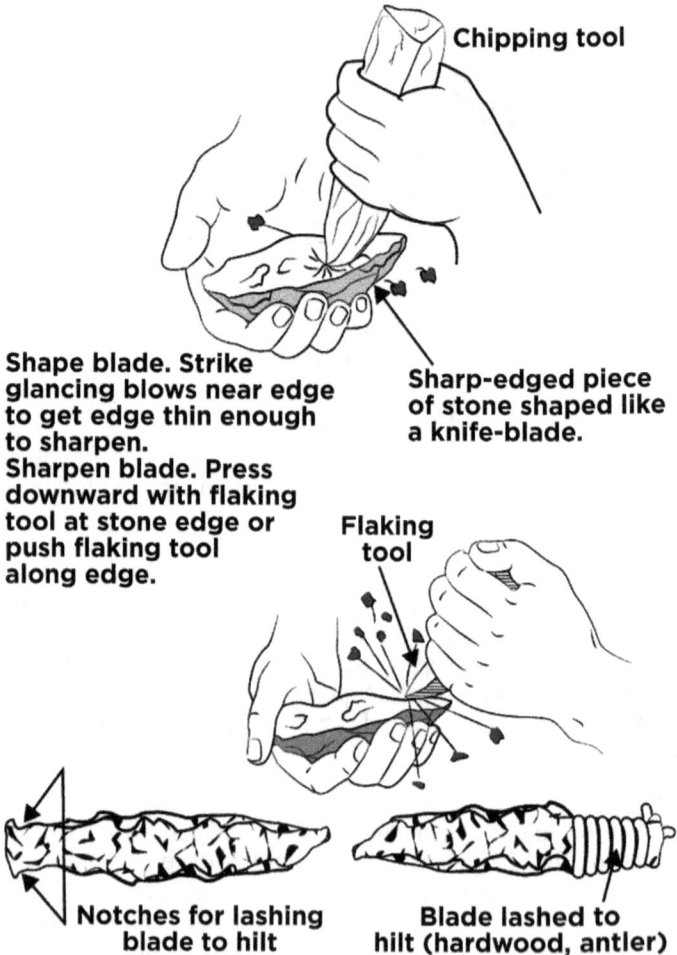

Chipping tool

1 **Shape blade. Strike glancing blows near edge to get edge thin enough to sharpen.**

Sharp-edged piece of stone shaped like a knife-blade.

2 **Sharpen blade. Press downward with flaking tool at stone edge or push flaking tool along edge.**

Flaking tool

Notches for lashing blade to hilt

Blade lashed to hilt (hardwood, antler)

Bone

8-19. Bone can be used as an effective field-expedient edged weapon when used only to puncture. It will not hold an edge and it may flake or break if used differently. To make a bone blade, begin by selecting a suitable bone. Larger bones, such as the leg bone of a deer or another medium-sized animal, are best. Lay the bone upon another hard object. Shatter the bone by hitting it with a heavy object, such as a rock. From the pieces, select a suitable pointed splinter. Further shape and sharpen this splinter by rubbing it on a rough-surfaced rock. If the piece is too small to handle, add a handle to it. Select a suitable piece of hardwood for a handle and lash the bone splinter securely to it.

Wood

8-20. Field-expedient edged weapons made from wood should be used only to puncture. Bamboo is the only wood that will hold an edge suitable for cutting. To make an edged tool from wood, begin by selecting a straight-grained piece of hardwood that is about 12 inches long and 1 inch in diameter. Fashion the blade about 6 inches long, and shave it down to a point. Use only the straight-grained portions of the wood. Do not use the core or pith, as it makes a weak point. Harden the point by a process known as fire hardening. If a fire is possible, dry the blade portion over the fire slowly until lightly charred; the drier the wood, the harder the point. After lightly charring the blade portion, sharpen it on a coarse stone. If using bamboo and after fashioning the blade, remove any other wood to make the blade thinner from the inside portion of the bamboo. Removal is done this way because bamboo's hardest part is its outer layer: keep as much of this layer as possible to ensure the hardest blade possible. When charring bamboo over a fire, char only the inside wood; do not char the outside.

Metal

8-21. Metal is best for making field-expedient edged weapons. A metal blade, when properly designed, can fulfill a knife's three uses—puncturing, slashing or chopping, and cutting. To make an edged weapon from metal, begin by selecting a suitable piece of metal, one that most resembles the desired product. Rub the metal on a rough-surfaced stone to obtain a point and cutting edge. Use a suitable flat, hard surface as an anvil and a smaller, harder object of stone or metal to hammer out the edge. If the metal is soft enough, an edge can be hammered out while the metal is cold. Make a knife handle from wood, bone, or other material that will protect your hand.

Other Materials

8-22. Other materials can be used to produce edged weapons. Glass is a good alternative to an edged weapon or tool if no other material is available. Obtain a suitable piece in the same manner as described for bone. Glass has a natural edge but is less durable for heavy work. Plastic can also be sharpened—if it is thick enough or hard enough—into a durable point for puncturing.

Spear Blades

8-23. To make a spear blade, use the same procedures to make the blade that are used to make a knife blade. Then select a shaft (a straight sapling) 4 to 5 feet long. The length should allow the ability to handle the spear easily and effectively. Attach the spear blade to the shaft using lashing. The preferred method is to split the handle, insert the blade, then wrap or lash it tightly (see Figure 8-4).

Figure 8-4. Bamboo for Spears

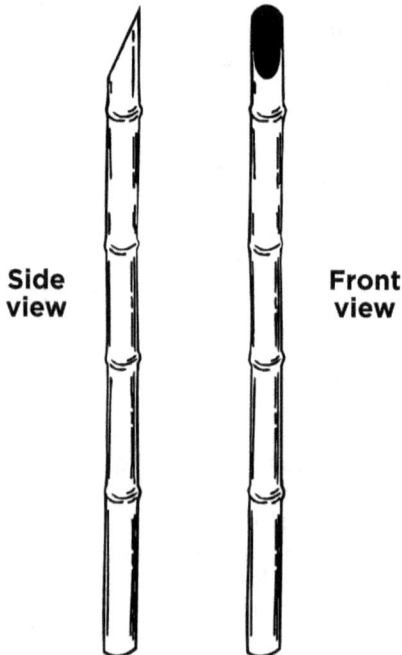

Side view

Front view

8-24. To make a spear without adding a blade, select a 4- to 5-foot long straight hardwood shaft and shave one end to a point. If possible, fire-harden the point. Bamboo makes an excellent spear. Starting 3 to 4 inches back from the end used as the point, shave down the end at a 45-degree angle. Remember, to sharpen the edges, shave only the inner portion.

Arrow Points

8-25. To make an arrow point, use the same procedures as for making a stone knife blade. Chert, obsidian, flint, and shell-type stones are best for arrow points; broken glass also makes an effective arrow point. You can fashion bone by flaking, like stone.

OTHER EXPEDIENT WEAPONS

8-26. Other field-expedient weapons can be made including the throwing stick, archery equipment, bola, sling, slingshot, and sap. The following Paragraphs explain how to make these weapons.

Throwing Stick

8-27. The throwing stick, commonly known as the rabbit stick, is very effective against small game such as squirrels, chipmunks, and rabbits. However, you must practice the throwing technique to develop accuracy and speed. The rabbit stick itself is a blunt stick, naturally curved at about a 45-degree angle (see Figure 8-5).

8-28. Select a stick with the desired angle from heavy hardwood such as oak. Shave off two opposite sides so that the stick is flat like a boomerang. To use the throwing stick, first align the target by extending the non-throwing arm in line with the mid- to lower-section of the target. Slowly and repeatedly raise the throwing arm up and back until the throwing stick crosses the back at about a 45-degree angle or is in line with the non-throwing hip. Bring the throwing arm forward until it is slightly above and parallel to the non-throwing arm. This will be the throwing stick's release point. This method should be practiced repeatedly to attain accuracy.

Figure 8-5. Throwing Stick

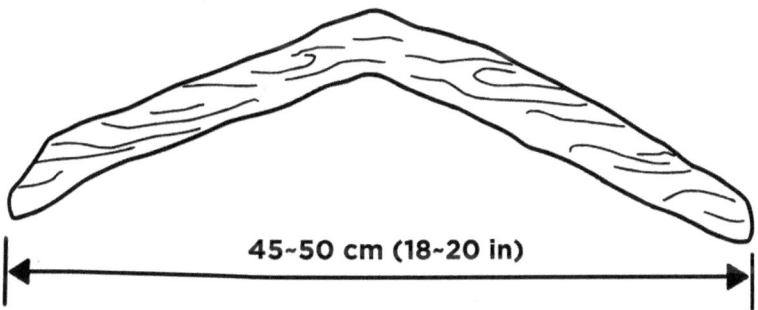

45~50 cm (18~20 in)

Archery Equipment

8-29. To make a bow, select a hardwood stick about three feet long that is free of knots or limbs. Carefully scrape the large end down until it has the same pull as the small end. Careful examination will show the natural curve of the stick. Always scrape from the side that faces you or the bow will break the first time it is pulled (see Figure 8-6 on page 267). Dead, dry wood is preferable to green wood. To increase the pull, lash a second bow to the first, front to front, forming an "X" when viewed from the side. Attach the tips of the bows with cordage and only use a bowstring on one bow.

8-30. Select arrows from the straightest dry sticks available. The arrows should be about half as long as the bow—and long enough that you can draw the bow to its full extent, maximizing its power, without the arrow falling off the bow. Scrape each shaft smooth all around. You will probably have to straighten the shaft. You can bend an arrow straight by heating the shaft over hot coals. Do not allow the shaft to scorch or burn. Hold the shaft straight until it cools.

8-31. You can make arrowheads from bone, glass, metal, or pieces of rock. You can also sharpen and fire-harden the end of the shaft. Fire hardening is actually a misnomer—you do not actually use fire, but the heat from coals. To fire-harden wood, hold it over hot coals or plunge it deep under the coals in the ashes, being careful not to burn or scorch the wood. The purpose of fire hardening is to harden the wood by drying the moisture out of it.

8-32. You must notch the ends of the arrows for the bowstring. Cut or file the notch; do not split it. Fletching (adding feathers to the notched end of an arrow) improves the arrow's flight characteristics. Fletching is recommended but not necessary on a field-expedient arrow.

8-33. While it may be relatively simple to make a bow and arrow, it requires considerable practice to attain proficiency at hitting a target. Choose a target of a size resembling the game you are most likely to encounter (of a material such as soft, dead wood that will not damage your arrows), and practice from a variety of distances to become familiar with the degree to which you will need to raise your aiming point at greater range.

Figure 8-6. Archery Equipment (Bow)

Shaping the bow

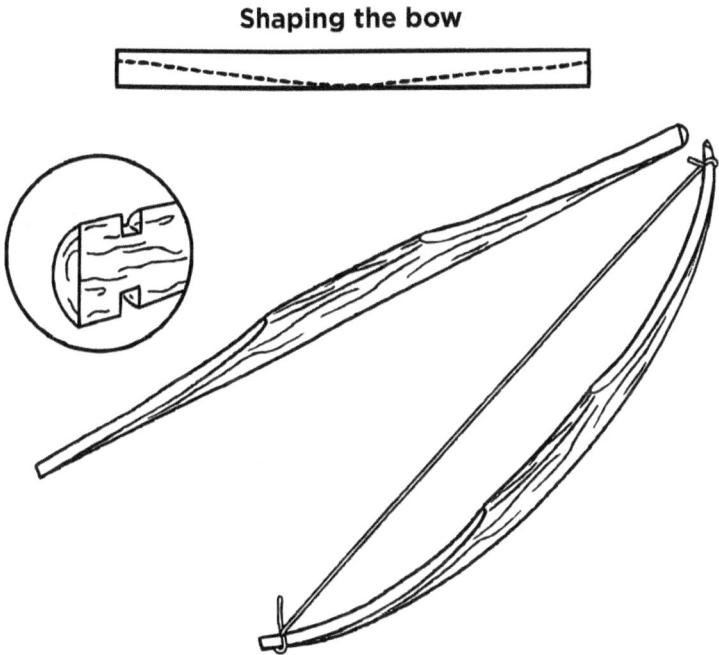

Bola

8-34. The bola is another field-expedient weapon that is easy to make (see Figure 8-7 on page 268). It is especially effective for capturing running game or low-flying fowl in a flock. To use the bola, hold it by the center knot and twirl it above your head, then release the knot so that the bola flies toward the target. When the bola is released, the weighted cords will separate from one another, held together at the center knot. These cords will wrap around and immobilize the fowl or animal it hits.

Figure 8-7. Bola

1 Use overhand knot to join three 60-cm (24-inch) cords.

2 Tie 0.25-kg (8-ounce) weight securely to ends of cords.

3 Hold by center knot and twirl the bola over your head. Release toward target.

Sling

8-35. Make a sling by tying two pieces of cordage, each about 60 centimeters (24 inches) long, at opposite ends of a palm-sized piece of leather or cloth. Place a rock in the cloth, wrap one cord around your middle finger, and hold in your palm. Hold the other cord between your forefingers and thumb. To throw the rock, spin the sling several times in a circle and release the cord between your thumb and forefinger. Practice to gain proficiency. The sling is very effective against small game.

Slingshot

8-36. To make a slingshot, find a Y in a green tree branch, then take some surgical tubing[1] and tie it off at the tips of the Y. Then use a small piece of material to make the ammo seat and tie it off to the other end of the surgical tubing. Small rocks can be used for slingshot ammunition.

Sap

8-37. A sap (or blackjack) can be made by filling a sock part way with rocks or other heavy material (metal, batteries, padlock, etc.) Use the sap like a weighted club.

---

1. If surgical tubing is unavailable, you can also use the rubber inner-tube from a bicycle or other pneumatic tire.

CORDAGE AND LASHING

8-38. Many materials are strong enough to use as cordage and lashing, and a number of natural and man-made materials are available in a survival situation. For example, you can make a cotton web belt much more useful by unraveling it. You can then use the string for other purposes such as fishing line, thread for sewing, and lashing. You can also use several strands of parachute cord and twist or braid them together to make a rope (see Figure 8-8). Bedsheets and other similar pieces of large fabric can be used in the same fashion. Parachute cord can also be broken down into its components of the outer cover (which works well for lashing) and the inner core strands (good for fishing line and sewing thread).

Figure 8-8. Making Cordage

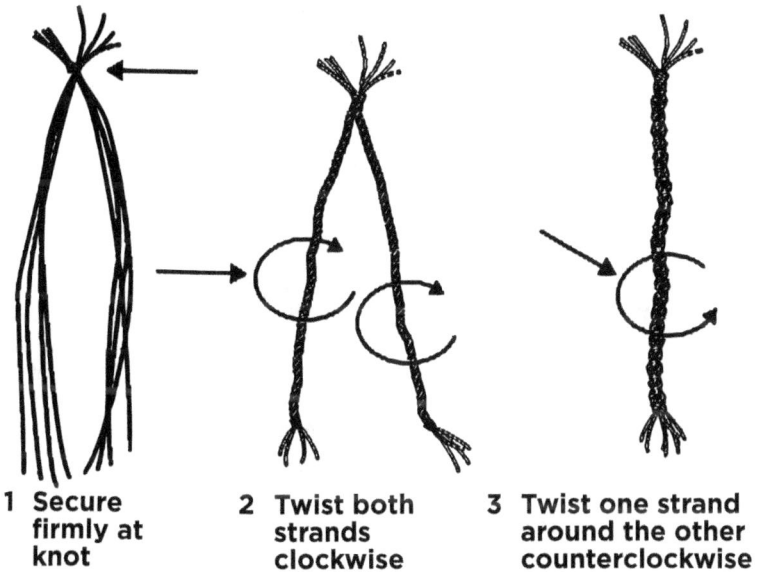

1 **Secure firmly at knot**

2 **Twist both strands clockwise**

3 **Twist one strand around the other counterclockwise**

NATURAL CORDAGE SELECTION

8-39. Before making cordage, conduct a few simple tests to determine the material's suitability. First, pull on a length of the material to test for strength. Next, twist it between your fingers and roll the fibers together. If it withstands this handling and does not snap apart, then tie an overhand knot with the fibers and gently tighten. If the knot does not break, it will work.

Lashing Material

8-40. Sinew is the best natural material for lashing small objects. Sinew can be obtained from the tendons of large game such as deer.

Remove the tendons from the game and dry them completely. Smash the dried tendons so that they separate into fibers. Moisten the fibers and twist them into a continuous strand. If stronger lashing material is needed, the strands can be braided. When sinew is used for small lashings, the lashings do not need knots as the moistened sinew is sticky and it hardens when dry. Another lashing material can be made with fibers from the inner bark of some trees, such as the linden, elm, hickory, white oak, mulberry, chestnut, and red and white cedar. Use these fibers to make cord. You should always test the cord produced to be sure it is strong enough for its intended purpose. These materials can be made stronger by braiding several strands together.

8-41. Rawhide can be used for larger lashing jobs. Make rawhide from the skins of medium or large game. After skinning the animal, remove any excess fat and any pieces of meat from the skin. Dry the skin completely. It does not need to be stretched as long as there are no folds to trap moisture, and the hair does not have to be removed from the skin. Cut the skin while it is dry. Make cuts about 1/4 inch wide. Start from the center of the hide and make one continuous circular cut, working clockwise to the hide's outer edge. Soak the rawhide for 2~4 hours or until it is soft. Use it wet, stretching it as much as possible while applying it. It will be strong and durable when it dries.

Rucksack Construction

8-42. The materials for constructing a rucksack or pack are almost limitless. Wood, bamboo, rope, plant fiber, clothing, animal skins, canvas, ponchos, evasion charts, and many other natural and man-made materials can be used to make a pack. There are several construction techniques for rucksacks. Many are very elaborate, but those that are simple and easy are often the most readily-made and reliable in a survival situation. Two types of simple-to-construct packs are the—

- **Horseshoe pack**. This pack is simple to make and use and relatively comfortable to carry over one shoulder. Lay available square-shaped material, such as a poncho, blanket or canvas, flat on the ground. Lay the items to be carried on one edge of the material. Pad the hard items. Roll the material (with the items) toward the opposite edge and tie both ends securely. Add extra ties along the length of the bundle. The pack can be draped over one shoulder with a line connecting the two ends (see Figure 8-9 on page 271).

Figure 8-9. Horseshoe Pack

- **Square pack**. This pack is easy to construct if rope or cordage are available (if not, make the cordage first). To make this pack, construct a square, box-shaped frame from bamboo, limbs, or sticks. Size will vary for each person and the amount of equipment carried (see Figure 8-10).

Figure 8-10. Square Pack

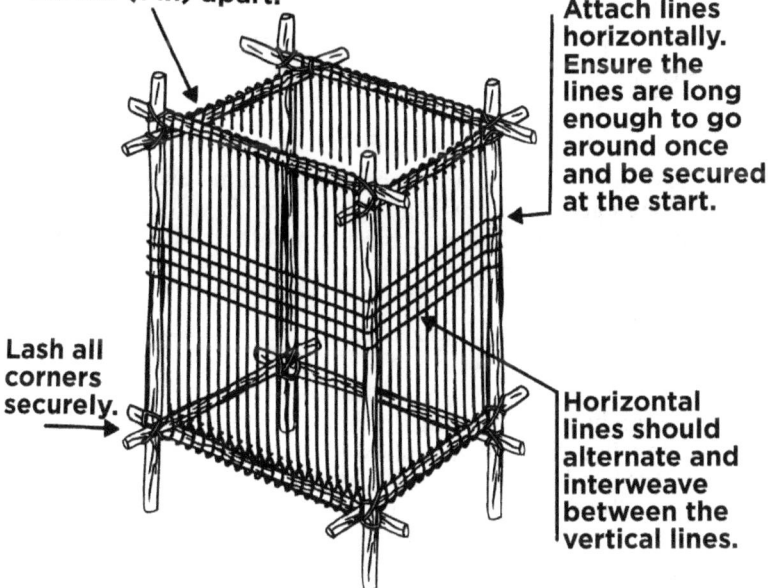

Attach lines or cordage all around, spaced about 2.5 cm (1 in) apart.

Attach lines horizontally. Ensure the lines are long enough to go around once and be secured at the start.

Lash all corners securely.

Horizontal lines should alternate and interweave between the vertical lines.

COOKING AND EATING UTENSILS

8-43. You can use many materials to make equipment for cooking, eating, and storing food. Usually all materials can serve some purpose when in a survival situation.

Bowls

8-44. Use wood, bone, horn, bark, or other similar material to make bowls. To make wooden bowls, use a hollowed out piece of wood that will hold food and enough water to cook. Hang the wooden container over the fire and add hot rocks to the water and food. Remove the rocks as they cool and add more hot rocks until the food is cooked.

CAUTION: Do not use rocks with air pockets, such as limestone and sandstone. They may explode while heating in the fire.

8-45. This method can also be used with containers made of bark or leaves. However, these containers will burn above the waterline unless they are kept moist or the fire is kept low. A section of bamboo also works very well for cooking. Cut out a section around two sealed joints. Cut out an opening for the food (the section you cut out can be kept and used as a lid, if helpful) and, after trimming their lower parts away, use the left and right overhanging sections to support the bamboo section over your fire on two stakes or bamboo sections (see Figure 8-11 on page 273).

CAUTION: Never heat a sealed section of bamboo. It will explode if heated because of trapped air and water in the section.

Forks, Knives, Spoons and Chopsticks

8-46. Carve forks, knives, and spoons from non-resinous woods so that you do not get a wood resin aftertaste and do not taint your food. Non-resinous woods include oak, birch, and other hardwood trees.

Note: Do not use trees that secrete a syrup or resin-like liquid on the bark or when cut.

Pots

8-47. Pots can be made from turtle shells or wood. As described with bowls, using hot rocks in a hollowed out piece of wood is very effective. Bamboo is the best wood for making cooking containers. To use turtle shells, first thoroughly boil the upper portion of the shell, then use it to heat food and water over a flame (see Figure 8-11 on page 273).

Figure 8-11. Containers for Boiling Food

Turtle shell

Coconut shell

Sea shell

Bamboo section

Improvising Pottery

8-48. Pottery has been used in most all areas of the world for storage vessels and cooking pots, plates and bowls. The process for making pottery in a survival situation involves finding a source of clay. Clay is formed through the weathering of granite rock. Try to find clay that is as pure as you can get. The best clay is found near the bends of streams and rivers, where the finest clay particles usually settle. Dig up a portion of clay to be used in the project, then leave it out to dry. Once dry, turn it into a powder and remove any impurities by sifting, as with flour (use your fingers if no sieve-like implement is available).

8-49. Once the clay is sifted and turned into a fine powder you should add in some tempering material to strengthen it. Items for tempering clay include ground up seashells, sand, ground up eggshells and other crushed-up pottery that has already been fired. Tempering reduces the plasticity of the clay, keeping it from shrinking unevenly and thus cracking during the firing process.

8-50. After the clay powder is tempered, add water to the mix and knead it to remove all of the air from the mixture. If any air is trapped in the clay, it will expand causing a break when it is fired.

8-51. Perform this process right before you are ready to form the clay. The proper consistency of clay is soft enough to form, but firm enough that it will not stick to the fingers. Once the clay reaches the desired consistency you can make it into plates, bowls, pots, and other containers. Once the clay is formed into shape leave it out to dry for several days. Once the clay project is dry rub it with a stone to close any open pores, sealing it and making it watertight (this is known as "burnishing" the clay).

8-52. Firing dried pottery items makes them more durable. For an improvised clay project made in the field the temperatures will be very low in comparison with those of a commercial kiln, so the items will likely deteriorate over time. Nonetheless, you should get plenty of use from them, and they are easily replaced. To fire the clay cover the item with a thick layer of soil and build a fire on top of it, or build a fire around the outer edge of the item. Ensure not to let the flames touch the project at first and that it heats up slowly over time. Move the fire in closer until the object is immersed in the fire and cooking at a high temperature for three to four hours. Take the item out after it has cooled off, clean it, and use it as necessary for cooking, water, and food storage.

Water Containers

8-53. Make water bottles from the stomachs of larger animals. Thoroughly flush the stomach out with water, and then tie off the bottom. Leave the top open, with some means of fastening it closed. Other items that can be used to hold water include non-lubricated prophylactics (placed inside a sock to protect the bladder) and plastic bags.

Weaving Baskets

8-54. Weave baskets to store materials like food, water, or anything else requiring organization, protection, and/or portability. Baskets can also make great fish and snake traps (see page 160). Being loosely-woven, baskets are excellent for storing food that needs air to prevent spoiling, like wild plants and berries. You can use many different materials to weave baskets, as long as they are flexible. Materials like shoots from a willow or hazel tree, pine or spruce roots, as well as different types of plants and grasses that can be made into cordage and then woven.

8-55. To make a basket, first determine the size of basket required based on the size of the objects it is to contain. Start with six flexible sticks cut to size to make the ribs or framework of the basket. Take the six ribs and divide them into two bundles of three sticks each. Make a cross of one bundle on top of the other. Secure the center of the cross with the material the basket is to be weaved out of with three strands of weaving material and a knot. Creating the base of the basket involves spreading the cross into six ribs and alternately weaving the three strands of material up and over one rib then under the next rib all the way around the circle until the base is the desired size. Once this is complete, the sides of the basket can be made by bending the ribs up. Making a small vertical slit in the rib before bending will keep it flexible and keep it from breaking.

8-56. The sides of the basket are constructed by continuing to alternate the pattern using the three strands. If a strand of material runs short or breaks, you can splice in a new piece of material by making a small twist from the end of the old piece to the end of the new piece and weave it in the same pattern and direction as the old strand. Once the basket has reached the desired height, tuck in the ends of the weaving material and trim the ends of the basket ribs to size.

APPENDIX A

SURVIVAL KNOTS AND ROPE

To be able to construct shelters, traps and snares, weapons and tools, and operate personnel recovery devices and equipment the isolated person should have a basic knowledge of ropes and knots and some of the terminology used with them.

ROPE TERMINOLOGY

A-1/A-2.[1] The following information is general terminology used for ropes and knots in this text (see Figure A-1 on page 278):

- **Bend**. A bend is a type of knot used to fasten two ropes together or to fasten a rope to a ring or loop.
- **Bight**. A bight is a U-shaped curve in a rope.
- **Dressing**. "Dressing" the knot refers to the orientation of all knot parts so that they are properly aligned, straightened, or bundled. Neglecting this can result in a 50 percent reduction in knot strength. This term is sometimes used for "setting" the knot, which involves tightening all parts of the knot so they bind on one another and make the knot operational. A loosely-tied knot can easily deform under strain and change, becoming a slipknot or worse, untying.
- **Fraps**. A means of tightening the lashings by looping the rope perpendicularly around the wraps that hold the spars or sticks together.

---

1. Paragraphs A-1 and A-2 have been combined for ease of use.

- **Hitch**. A type of knot used to tie a rope around a timber, pipe, or post so that it will hold temporarily, but can readily be untied.
- **Lashings**. A means of using wraps and fraps to tie two or three spars or sticks together to form solid corners or to construct tripods. Lashings begin and end with clove hitches.
- **Lay**. The "lay" of the rope is the same as the twist of the rope.
- **Line**. A single thread, string, or cord.

Note: In naval terminology, the word "line" is used in general to refer to any fabric rope (as opposed to wire rope), or more specifically any piece of rope that has been cut for a specific purpose (such as a "lifeline").

- **Loop**. A fold or doubling of the rope through which another rope can be passed.
- **Overhead turn** or **overhead loop**. An overhead loop is made when the running end passes over the standing part.
- **Pig tail**. That part of the running end that is left after tying the knot. It should be no more than 4 inches long to conserve rope and prevent interference.
- **Rope**. A rope is made of strands of fiber twisted or braided together.
- **Round turn**. The same as a turn, with the running end leaving the circle in the same general direction as the standing part.
- **Running end**. The running end is the free or working end of a rope.
- **Standing end**. The standing end is the static part of the rope, or the rest of the rope, excluding the running end.
- **Turn**. A loop around an object such as a post, rail, or ring with the running end continuing in the opposite direction from the standing end.
- **Underhand turn** or **overhead loop**. An underhand turn or loop is made when the running end passes under the standing part.
- **Whipping**. Any method of preventing the strands at the end of a rope from untwisting or becoming unwound. It is done by wrapping the end tightly with a small cord, tape or other means. Before cutting a rope, it should be done on both sides of the anticipated cut: this prevents the rope from immediately untwisting.
- **Wraps**. Simple wraps of rope around two poles or sticks (square lashing) or three poles or sticks (tripod lashing). Wraps begin and end with clove hitches and get tighter with fraps. All together, they form a lashing.

Figure A-1. Elements of Ropes and Knots

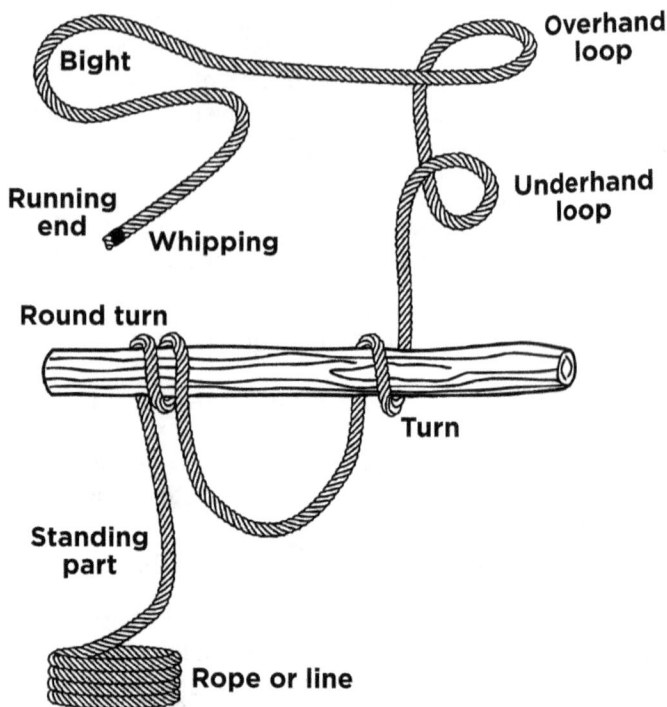

BASIC KNOTS

A-3. Descriptions of basic knots and the methods for tying them are discussed in Paragraphs A-4~A-25.

OVERHAND KNOT

A-4. This is the most commonly used and the simplest of all knots, which most people tie every day as the first part of tying their shoelaces. An overhand knot may be used to prevent the end of a rope from untwisting, to form a knot at the end of a rope, or as part of another knot. This knot should replace the half-hitch as a finishing knot for other knots. This knot alone will reduce the strength of a straight rope by 55 percent. To tie an overhand knot, make a loop near the end of the rope and pass the running end through the loop, pulling it tight (see Figure A-2 on page 279).

Figure A-2. Overhand

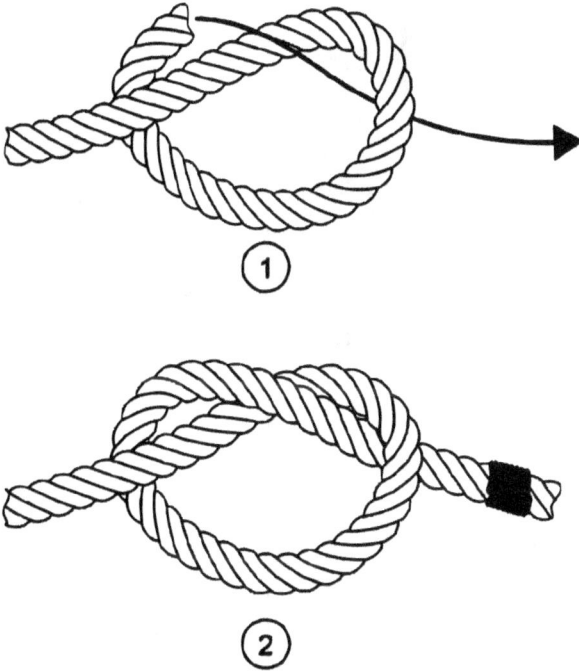

ROUND TURN AND TWO HALF HITCHES

A-5. The round turn and two half hitches (see Figure A-3 on page 280) is the main anchor knot for one-rope bridges and other applications when a good anchor knot is required and where high loads would make other knots jam and become difficult to untie. It is most often used to anchor rope to a pole or tree.

Figure A-3. Round Turn and Two Half Hitches

FIGURE-EIGHT KNOT

A-6. Use the figure-eight knot to form a larger knot than would be formed by an overhand knot at the end of a rope. A figure-eight knot is used at the end of a rope to prevent the ends from slipping through a fastening or loop in another rope. To make a figure-eight knot, make a loop in the standing part, pass the running end around the standing part back over one side of the loop, and down through the loop. The running end can then be pulled tight (see Figure A-4 on page 281).

Figure A-4. Figure-eight Knot

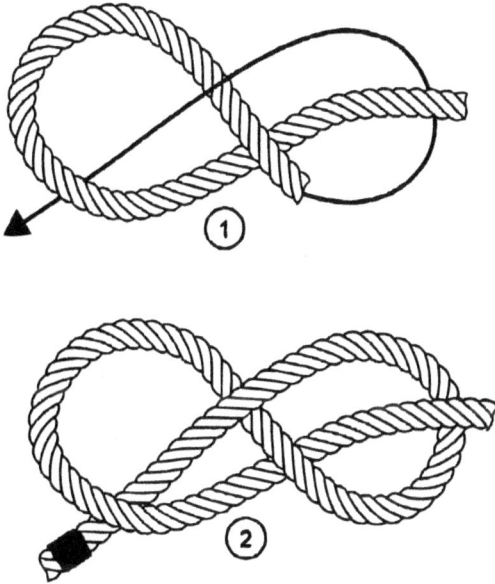

KNOTS FOR JOINING ROPE

A-7. When you require rope longer than the individual pieces you are equipped with or have improvised, you can securely tie multiple ropes together using knots designed for adjoining rope of equal and unequal diameters in wet and dry conditions. Knots for joining two ropes fall into the following categories:

- **Square knot.**
- **Single sheet bend.**
- **Double sheet bend.**

SQUARE KNOT

A-8. The square knot (also known as the reef knot) is used for tying two ropes of equal diameter together to prevent slippage (see Figure A-5 on page 282). To tie a square knot, lay the running end of each rope together but pointing in opposite directions. The running end of one rope can be passed under the standing part of the other rope. Bring the two running ends up away from the point where they cross and cross them again. Once each running end is parallel to its own standing part, the two ends can be pulled tight. If each running end does not come parallel to the standing part of its own rope, the knot is called a "granny knot." Because it will slip under the strain, a granny knot should not be used.

A-9. A square knot can also be tied by making a bight in the end of one rope and feeding the end of the other rope through and around this bight. The running end of the second rope is routed from the standing side of the bight. If the procedure is reversed, the resulting knot will have a running end parallel to each standing part but the two running ends will not be opposite each other. This knot is called a "thief" knot. It will slip under strain and is difficult to untie. A true square knot will draw tighter under strain. A square knot can be untied easily by grasping the bends of the two bights and pulling the knot apart.

Figure A-5. Square Knot

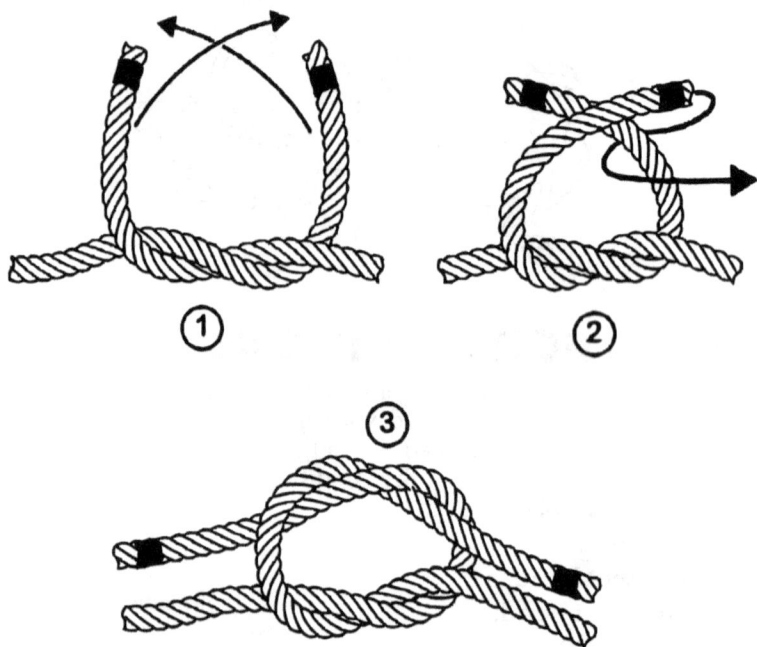

SINGLE SHEET BEND

A-10. The single sheet bend, sometimes called a weaver's knot, is used for tying together two dry ropes of unequal size. To tie the single sheet bend, the running end of the smaller rope should pass through a bight in the larger rope. The running end should continue around both parts of the larger rope and back under the smaller rope. The running end can then be pulled tight. This knot will draw tight under light loads but may loosen or slip when the tension is released (see Figure A-6 on page 283).

Figure A-6. Single Sheet Bend

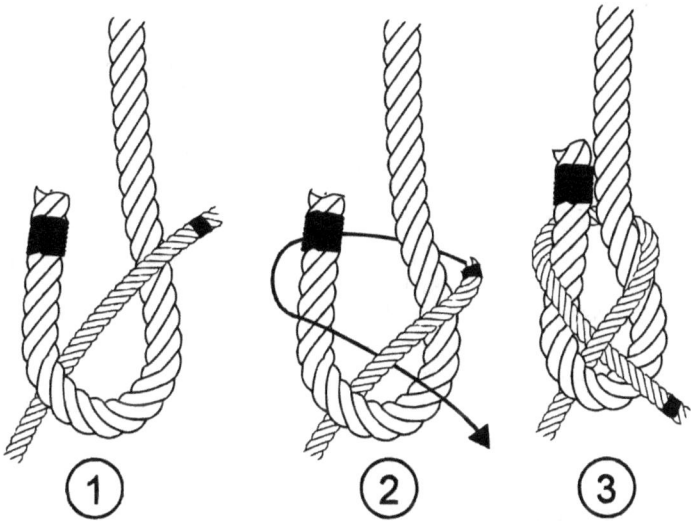

DOUBLE SHEET BEND

A-11. The double sheet bend works better than the single sheet bend for joining ropes of equal or unequal diameter, joining wet ropes, or tying a rope to an eye. It will not slip or draw tight under heavy loads. To tie a double sheet bend, a single sheet bend is tied first. However, the running end is not pulled tight. One extra turn is taken around both sides of the bight in the larger rope with the running end for the smaller rope. Then tighten the knot (see Figure A-7).

Figure A-7. Double Sheet Bend

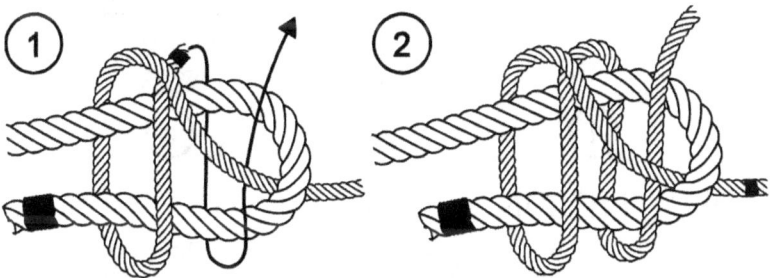

KNOTS FOR MAKING LOOPS

A-12. Knots for making loops fall into the following categories:
- **Bowline**.
- **Bowline on a bight**.

- **French bowline**.
- **Speir knot**.
- **Overhand knot fixed loop**.

Note: The knot you should learn and commit to memory for use if you are required to create a secure loop at the end of a rope for personnel recovery operations is the overhand fixed loop.

BOWLINE

A-13. The bowline is a useful knot for forming a loop in the end of a rope. It is also easy to untie. To tie a bowline knot, the running end of the rope passes through the object to be affixed to the bowline and forms a loop in the standing part of the rope. The running end is then passed through the loop from underneath and around the standing part of the rope, and back through the loop from the top. The running end passes down through the loop parallel to the rope coming up through the loop. The knot is then pulled tight (see Figure A-8).

Figure A-8. Bowline Knot

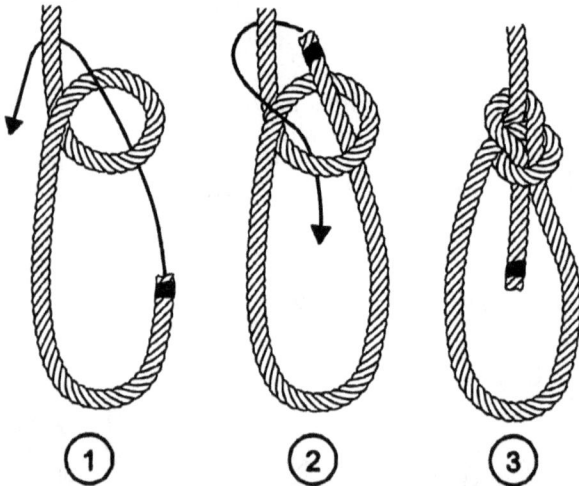

BOWLINE ON A BIGHT

A-14. It is sometimes desirable to form a loop at some point in a rope other than at the end. The bowline on a bight can be used for this purpose. It is easily untied and will not slip. The same knot can be tied at the end of a rope by doubling the rope for a short section. A doubled portion of the rope is used to form a loop as in the case of the bowline. The bight end of the doubled portion is passed up through the loop, back down up around the entire knot, and tightened (see Figure A-9 on page 285).

Figure A-9. Bowline on a Bight Knot

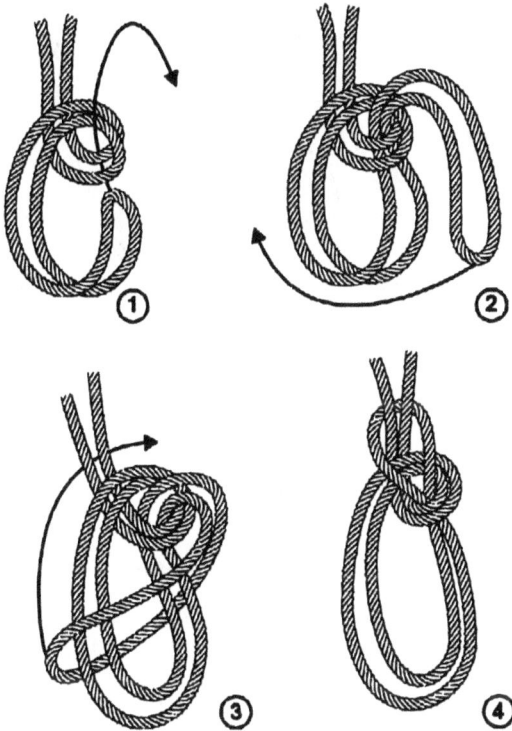

FRENCH BOWLINE

A-15. The French bowline is sometimes used as a sling for lifting people. When used in this manner, one loop is used as a seat and the other loop is put around the body under the arms. The weight of the injured person keeps the two loops tight so that the victim cannot fall out of the sling. Therefore, it is particularly useful as a sling for someone who is unconscious.

A-16. The French bowline is started in the same way as the simple bowline. Make a loop in the standing part of the rope. The running end is passed through the loop from underneath and a separate loop is made. The running end is passed through the loop, again from underneath, around the back of the standing part and back through the loop so that it comes out parallel to the looped portion. The standing part of the rope is pulled to tighten the knot, leaving two loops (see Figure A-10 on page 286).

Figure A-10. French Bowline

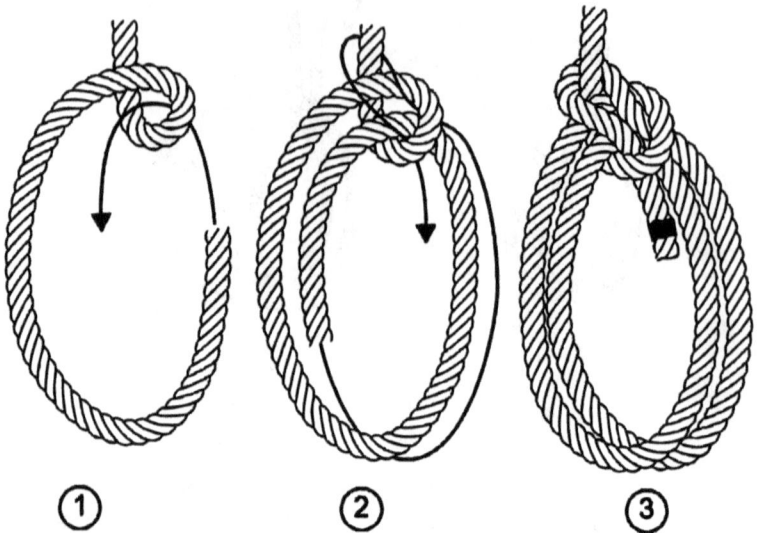

SPEIR KNOT

A-17. A speir knot is used when a fixed loop, a nonslip knot, and a quick release are required. It can be tied quickly and released by a pull on the running end. To tie a speir knot, the running end is passed through a ring or around a pipe or post and brought back on the left side of the standing part. Both hands are placed, palms up, under both parts of the rope with the left hand higher than the right hand; grasping the standing part with the left hand and the running end with the right hand. The left hand is moved to the left and the right hand to the right to form two bights. The left hand is twisted a half turn toward the body so that the bight is twisted into a loop. Pass the bight over the rope down through the loop. The speir knot is tightened by pulling on the bight and the standing part (see Figure A-11 on page 287).

Figure A-11. Speir Knot

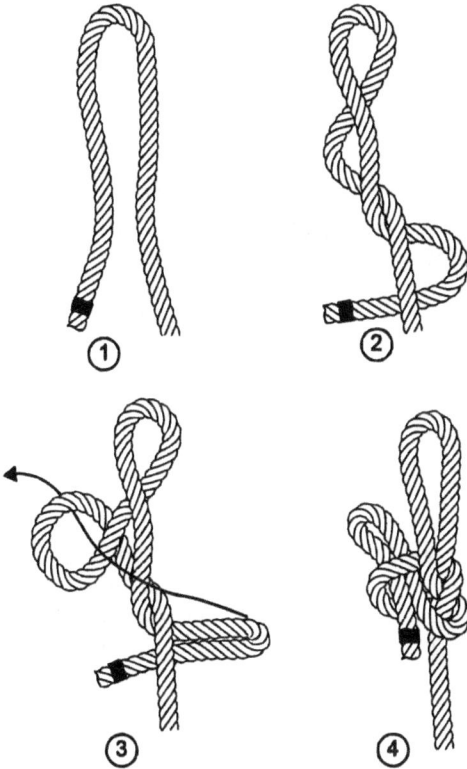

OVERHAND KNOT FIXED LOOP

A-18. The overhand knot fixed loop is a simple knot which forms a fixed loop in a rope. Make this knot by tying an overhand knot in a bight: create a bight of the desired size for your loop, then tie an overhand knot in the bight to create the loop (using both parts of the rope at once). This knot can be tied anywhere along a rope (does not need any working end). The knot can be used for attaching clips, hooks, other rope, etc. The disadvantage of this knot is that it is likely to jam tight when the rope has been pulled and the knot may need to be cut off. This knot is used in conjunction with a personnel recovery.

HITCHES

A-19. A hitch is any of various knots used to form a temporary noose in a rope or to secure a rope around a timber, pipe, post, or another rope, so that it will hold temporarily but can be readily undone or moved. The types of hitches discussed in this book are as follows:

- **Half hitch**.

- **Two half hitches**.
- **Timber hitch**.
- **Timber hitch and half hitch**.
- **Clove hitch**.
- **Sheepshank**.
- **Prusik**.

HALF HITCH

A-20. This is the simplest of all knots and used to be the safety, or finishing knot for all Army knots. Because it has a tendency to undo itself without load, it has since been replaced by the overhand (see page 278). The half hitch is used to tie a rope to a timber or to another larger rope. It is not a very secure knot or hitch and is used for temporarily securing the free end of a rope. To tie a half hitch, the rope is passed around the timber, bringing the running end around the standing part, and back under itself (see Figure A-12 on page 289).

TWO HALF HITCHES

A-21. A quick method for tying a rope to a timber or pole is to use two half hitches. The running end of the rope is passed around the pole or timber, and a turn is taken around the standing part and under the running end. This is a one half hitch. The running end is passed around the standing part of the rope and back under itself again (see Figure A-12 on page 289).

Figure A-12. Half Hitch and a Two Half Hitch

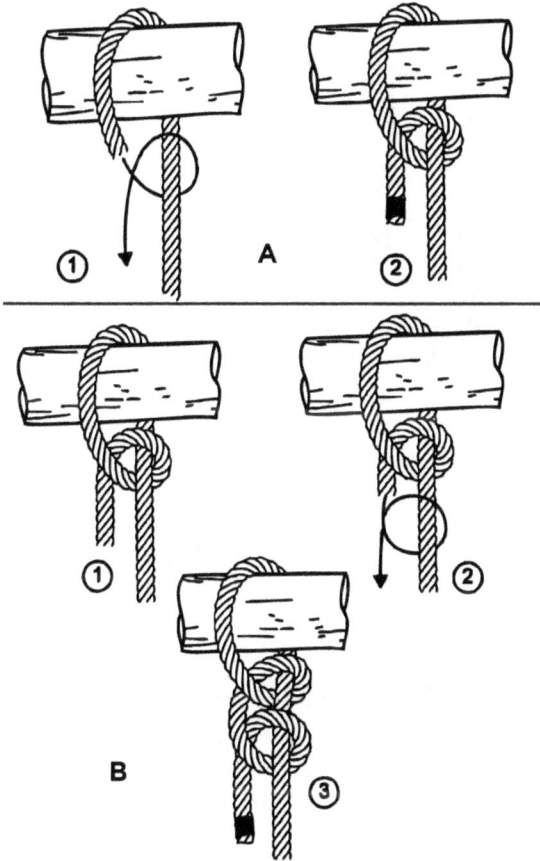

TIMBER HITCH

A-22. The timber hitch is used for moving heavy timbers or poles. To make the timber hitch, a half hitch is made and similarly the running end is turned about itself at least another time. These turns must be taken around the running end itself or the knot will not tighten against the pole (see Figure A-13 on page 290).

Figure A-13. Timber Hitch

① ②

TIMBER HITCH AND HALF HITCH

A-23. To get a tighter hold on heavy poles for lifting or dragging, a timber hitch and half hitch are combined. The running end is passed around the timber and back under the standing part to form a half hitch. Further along the timber, a timber hitch is tied with the running end. The strain will come on the half hitch and the timber hitch will prevent the half hitch from slipping (see Figure A-14).

Figure A-14. Timber Hitch and Half Hitch

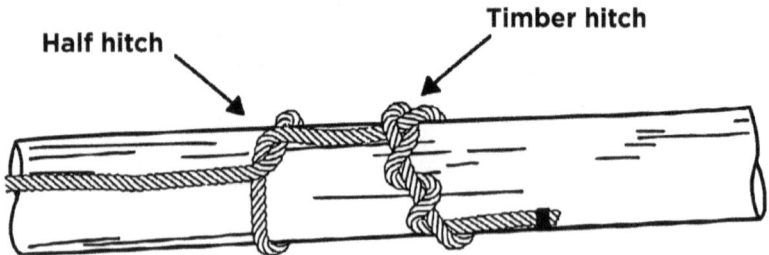

Half hitch   **Timber hitch**

CLOVE HITCH

A-24. A clove hitch is used to fasten a rope to a timber, pipe, or post. It can be tied at any point in a rope. To tie a clove hitch in the center of the rope, two turns are made in the rope close together. They are twisted so that the two loops lay back-to-back. These two loops are slipped over the timber or pipe to form the knot. To tie the clove hitch at the end of a rope, the rope is passed around the timber in two turns so that the first turn crosses the standing part and the running end comes up under itself on the second turn (see Figure A-15).

Figure A-15. Clove Hitch

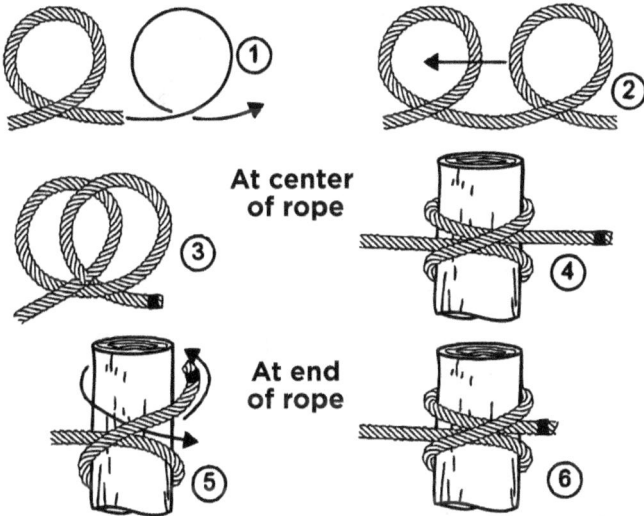

SHEEPSHANK

A-25. A sheepshank is a method of shortening a rope, but it may also be used to take the load off a weak spot in the rope. To make the sheepshank (which is never made at the end of a rope), two bights are made in the rope so that three parts of the rope are parallel. A half hitch is made in the standing part over the end of the bight at each end (see Figure A-16 on page 292).

Figure A-16. Sheepshank

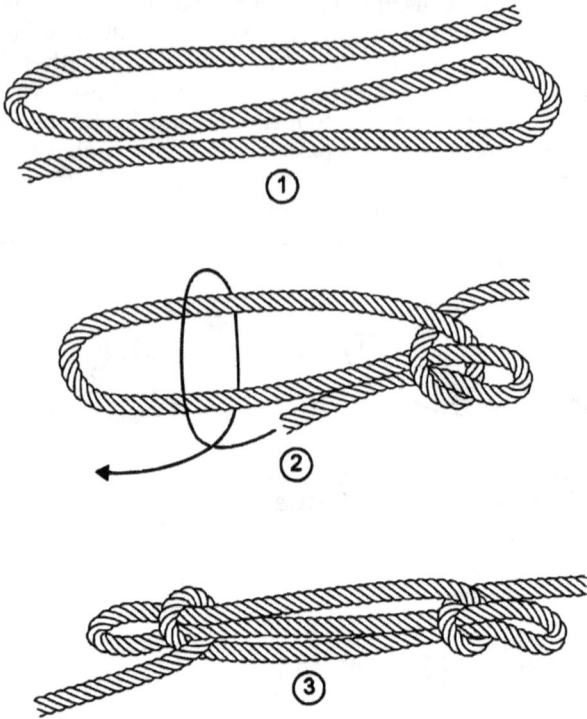

PRUSIK

A-26. This knot ties a short rope around a longer rope (for example, a sling rope around a climbing rope) in such a manner that the short rope will slide on the longer rope if no tension is applied, and will hold if tension is applied on the short rope. This knot can be tied with an end of rope or bight of rope. When tied with an end of rope, the knot is finished off with a bowline. The non-slip nature of the knot on another rope allows climbing of ropes using short ropes as foot holds. This knot can also be used to anchor ropes, or at the end of a traction splint on a branch or ski pole. (See Figure A-17 on page 293).

Figure A-17. Prusik

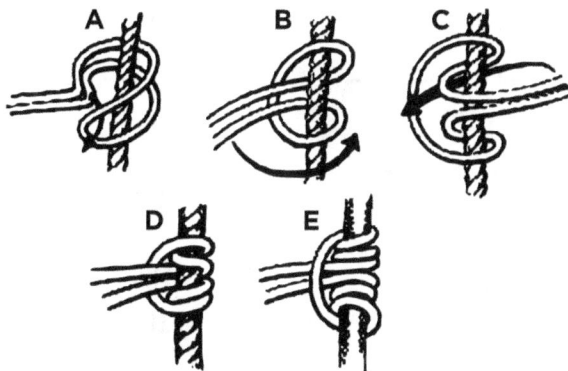

LASHINGS

A-27. A lashing is as rope, wire, or chain used for binding, wrapping, or fastening. Numerous items require lashings for construction; for example, shelters, equipment racks, and smoke generators. The following three types of lashings are discussed in this appendix:

- **Square lashing**.
- **Diagonal lashing**.
- **Shear lashing**.

SQUARE LASHING

A-28. The square lash is used to secure one pole at right angles to another pole.[1] Square lashing is started with a clove hitch around the log, immediately under the place where the crosspiece is to be located. In laying the turns, the rope goes on the outside of the previous turn around the crosspiece, and on the inside of the previous turn around the log. The rope should be kept tight. Three or four turns are necessary. Two or three "frapping" turns are made between the crosspieces. The rope is pulled tight; this will bind the crosspiece tightly together. It is finished with a clove hitch around the same piece on which the lashing was started (see Figure A-18 on page 294).

---

1. Another lash that can be used for the same purpose is the diagonal lash (see page 294).

Figure A-18. Square Lashing

1 **Start by tying a clove hitch to the vertical spar, just below where the horizontal spar will be.**

2 **Twist the end of the rope around the vertical part of the rope for a clean look, then wrap the rope around the horizontal and vertical spars, binding them together.**

3 **Continue by wrapping the rope three or four times around the vertical and horizontal spars.**

4 **Make two or three frapping turns between the spars, around the rope itself. Pull these frapping turns very taut. Finish by tying a clove hitch to the horizontal spar.**

DIAGONAL LASHING

A-29. Diagonal lashing is started with a clove hitch around the two poles at the point of crossing. Three turns are taken around the two poles. The turns lie beside each other, not on top of each other. Three more turns are made around the two poles, this time crosswise over the previous turns. The turns are pulled tight. A couple of frapping turns are made between the two poles, around the lashing turns, making sure they are tight. The lashing is finished with a clove hitch around the same pole the lash was started on (see Figure A-19 on page 295).

Figure A-19. Diagonal Lashing

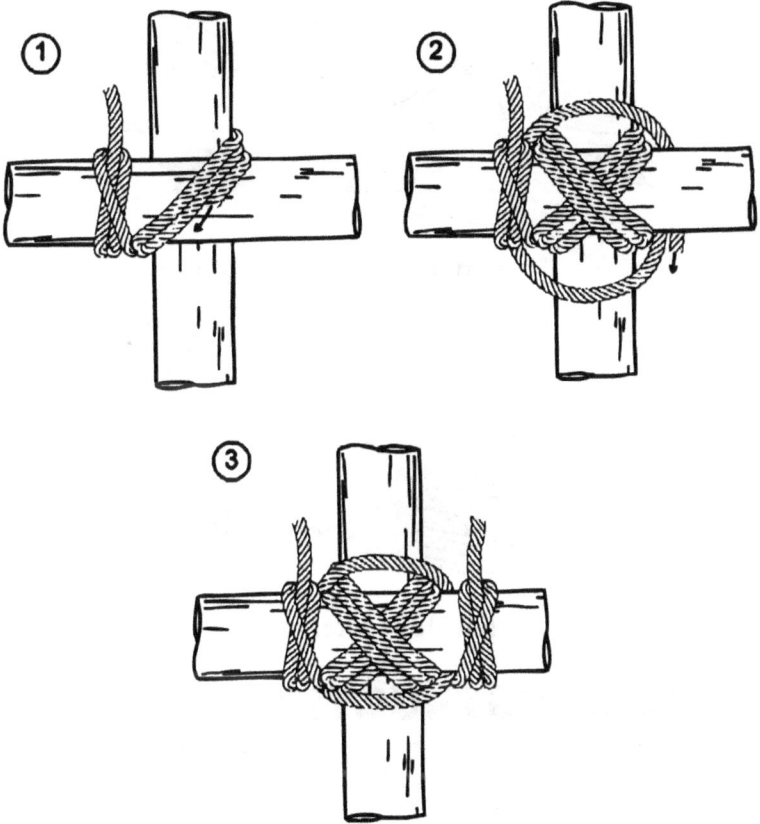

SHEAR LASHING

A-30. Shear lashing is used for lashing two or more poles in a series. The desired numbers of poles are placed parallel to each other and the lash is started with a clove hitch on an outer pole. The poles, laid loosely beside each other, are often lashed together using seven or eight turns of the rope. Make frapping turns between each pole. The lashing is finished with a clove hitch on the pole opposite that on which the lash was started (see Figure A-20 on page 296).

Figure A-20. Shear Lashing

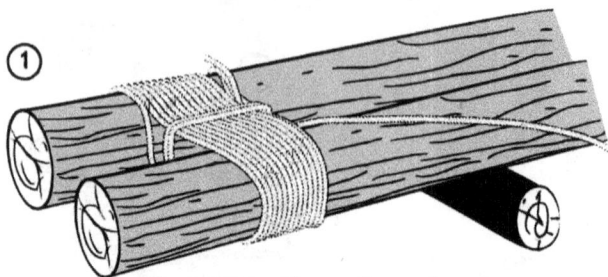

Tie two or three frapping turns.

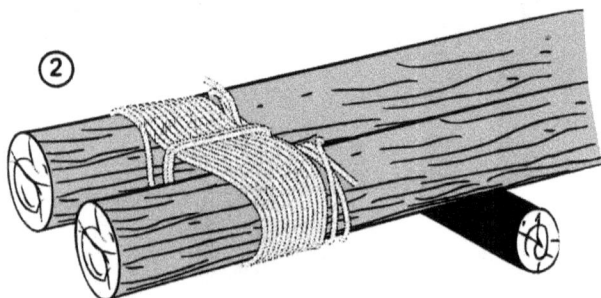

Finish with a clove hitch on opposite spar.

ROPE CONSTRUCTION

A-31. There are two basic types of rope to use in a survival situation—natural fiber ropes and man-made fiber ropes. Both types have different strengths and weaknesses. You should learn as much about the properties of rope as you can, before you find yourself in a survival situation. This will allow you to use or, in some cases, improvise ropes of your own; creating the right tool for the task that will keep you from wasting time, energy, and effort.

NATURAL FIBER TYPES

A-32. Natural fiber ropes are made from natural fibers from plants and animals such as grasses, sisal, hemp, cotton, manila, hair, leather, and sinew. These fibers are either twisted or braided into a single flexible structure that has a high degree of tensile strength. Natural ropes are resistant to friction-generated heat but can be destroyed if the fibers are cut with a sharp rock or knife. Natural rope can be cut into sections using abrasion; for example, rubbing it vigorously across rock, cement, concrete, metal, or wood surfaces. When the integrity of the fibers is damaged, the rope can be separated into sections or broken down to be used for other purposes. Most natural ropes are thicker in diameter and can therefore be difficult to tie knots with. If not cared for and kept clean, natural fiber ropes can begin to break down and lose tensile strength.

MANMADE FIBER TYPES

A-33. Man-made fiber ropes are manufactured from fibers made out of fabrics like polypropylene, nylon, and Kevlar. Man-made ropes are braided or twisted into a single, flexible structure. Man-made fiber ropes are strong because the fibers are strong and can be manufactured in one long continuous strand, with no splices of fiber being used during the construction process.

A-34. Man-made fiber ropes are water- and weather-resistant. Because these ropes are essentially made from plastic material, they are susceptible to break down from friction-generated heat and abrasion. Isolated personnel can use a piece of Kevlar-like material or parachute cord to saw these lengths of rope into usable pieces through melting them with friction-generated heat. These ropes can also be cut with a knife, a rock, or a piece of glass. Man-made fiber ropes are good for tying knots but should still be kept clean and out of the sun for UV protection and damage mitigation.

A-35. Kernmantle rope is a type of a man-made fiber rope that has an outer sheath with smaller fibers running through the center known as intercore fibers. This type of construction makes kernmantle ropes very strong and abrasion-resistant. They are commonly used for rock climbing and other functions like rappelling. Kernmantle ropes come in two types—static (will not stretch) and dynamic (designed to stretch and return to its original manufactured shape). Parachute cord is a good example of the static type.

CONSTRUCTION TECHNIQUES

A-36. Almost any natural fibrous material can be spun into good serviceable rope or cord, and many materials which have a length of 12~24 inches or more can be braided. Ropes up to 3 and 4 inches in diameter can be "laid" by four people, and tensile strength for bush-made rope of 1 inch in diameter ranges from 100 pounds to as high as 3,000 pounds.

A-37. Using a three-lay rope of 1-inch diameter as standard, the following table of tensile strengths (table A-1) may serve to illustrate general strengths of various materials. For safety's sake, the lowest figure should always be regarded as the tensile strength.

Table A-1. Tensile Strength of Natural Fibers

Fiber	Tensile Strength (in Pounds)
Green grass	100~250
Bark fiber	500~1,500
Palm fiber	650~2,000
Sedges	2,000~2,500
Money rope (lianas)	560~700
Lawyer vine (calamus), ½-inch diameter	1,200

Note: Doubling the diameter quadruples the tensile strength; halving the diameter reduces the tensile strength to one-fourth.

Twisting Technique

A-38. Use any material with long strong threads or fibers which have been previously tested for strength and pliability. Gather the fibers into loosely held strands of even thickness. Each of these strands is twisted clockwise. The twist will hold the fibers together. The strands should be formed into a one-eighth inch diameter. As a general rule, there should be about 15 to 30 fibers to a strand. Two, three, or four of these strands are later twisted together or "layered" with a counterclockwise twist, while at the same time, the separate strands which have not yet been laid up are twisted clockwise. Each strand must be of equal twist and thickness (see Figure A-21 on page 299).

Figure A-21. Twisting Fibers

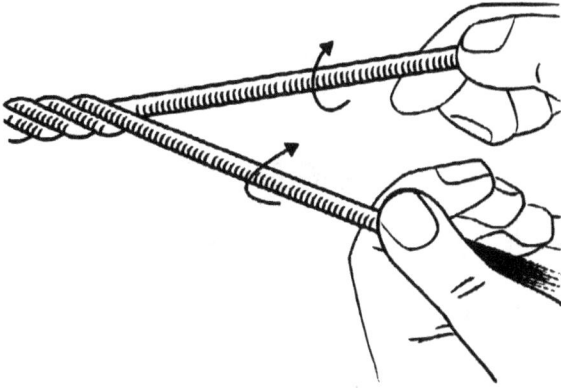

A-39. In a similar manner, the twisted strands are put together into lays, and the lays into ropes. Twist the strands together and ensure that the twisting is even, the strands are uniform, and the tension on each strand is equal. In "laying," care must be taken to ensure that each of the strands is evenly "laid up;" that is; one strand does not twist around the other.

A-40. When spinning fine cords for fishing lines and snares, considerable care must be taken to keep the strands uniform and the lay even. Fine thin cords of no more than 1/32-inch thickness can be spun with the fingers and are capable of taking a breaking strain of 20 to 30 pounds or more.

A-41. Normally two or more people are required to spin and lay up the strands for cord. However, many native peoples spin cord unaided. They twist the material by running the flat of the hand along the thigh, with the fibrous material between hand and thigh; with the free hand, they feed in fiber for the next "spin." Using this technique, one person can make long lengths of single strands. This method of making cord or rope with the fingers is slow if any considerable length of cord is required.

Braiding Techniques

A-42. One person may require a length of rope. If there is no help available to spin materials, it is necessary to find reasonably long material. With this material you can braid and make suitable rope without assistance. The usual three-strand braid makes a flat rope and, while quite good, it does not have finish or shape and is not as tight as the four-strand braid. On other occasions, it may be necessary to braid broad bands for belts or shoulder straps. There are many fancy braids which can be developed from these, but the following three braids are basic and essential for practical woodcraft work. A general rule for all braids is to work from the outside into the center.

Three-plait Braid

A-43. Use the following steps to make a three-plait braid (see Figure A-22):

Step 1 Pass the right-hand strand over the strand to the left.

Step 2 Pass the left-hand strand over the strand to the right.

Step 3 Repeat this process alternately from left to right.

Figure A-22. Three-strand Braid

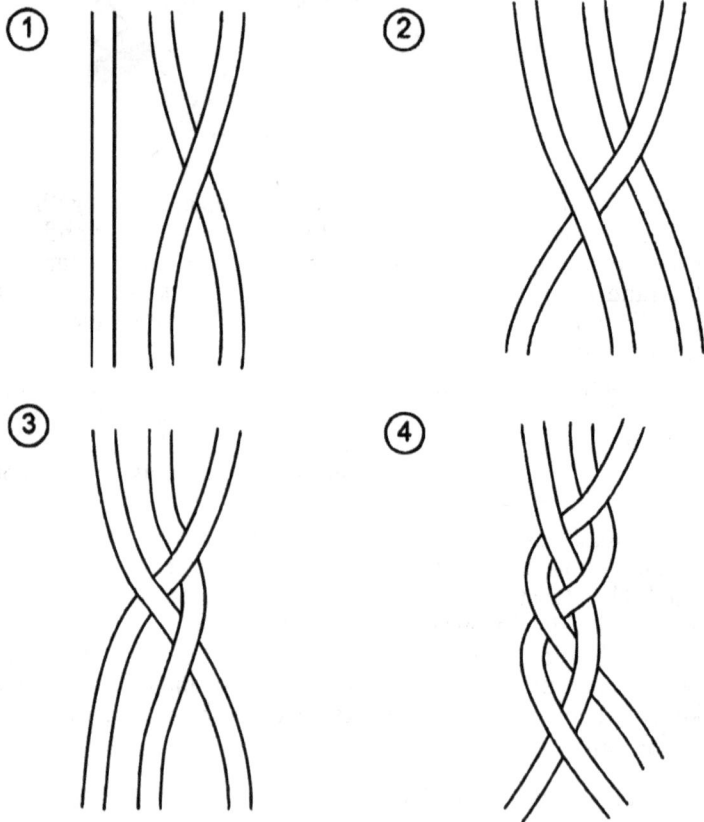

Broad Braid

A-44. Use the following steps to make a broad braid (see Figure A-23 on page 301):

Step 1 Hold six or more strands flat and together. Pass a strand in the center over the next strand to the left.

Step 2 Pass the second strand to the left of center toward the right and over the first strand so that it points toward the right.

Step 3 Take the strand to the right of the first one and weave it under and over the two strands to its left.

Step 4 Take the strand to the right of the one you just moved, and weave it over and under the two strands to its left.

Step 5 Weave the next strands from left and right, alternatively, towards the center. The finished braid should be tight and close.

Figure A-23. Broad Braid

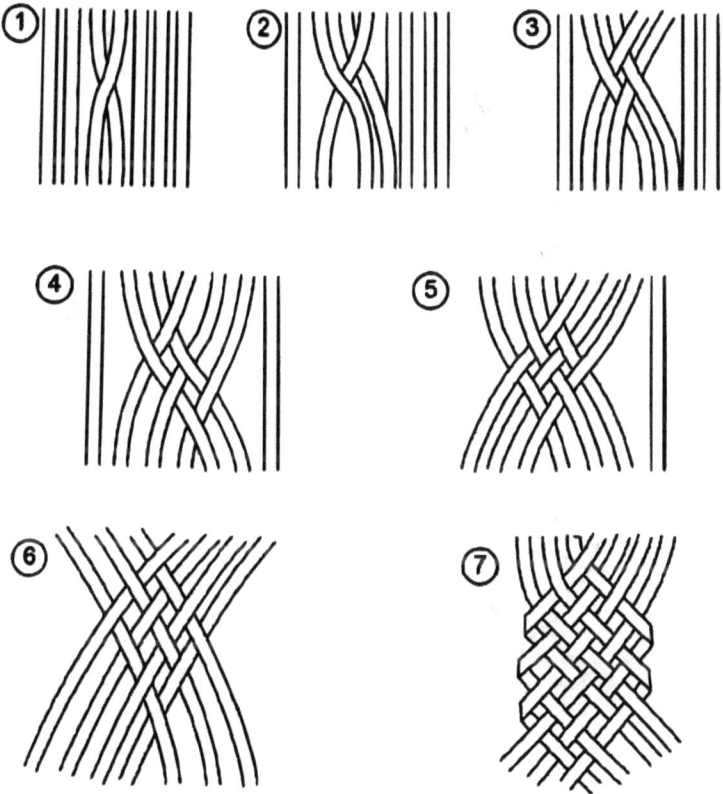

A-45. Use the following steps to finish the broad braid (see Figure A-24 on page 302):

Step 1 Lay one of the center strands back upon itself.

Step 2 Take the first strand which it enclosed in being folded back, and weave this back upon itself.

Step 3 Lay back the strand from the opposite side and weave it between the strands already braided. All the strands should be so woven back that no strands show an uneven pattern, and there should be a regular under-over-under of the alternating weaves. If the braid is tight, there may be a difficulty in working the loose ends between the plaited strands. This can be done easily by sharpening a thin piece of wood to a chisel edge to open the strands sufficiently to allow the ends being finished to pass between the woven strands.

Step 4 Roll the braid under a bottle or other round object to flatten it and achieve a smooth final finish.

Figure A-24. Finish Braid

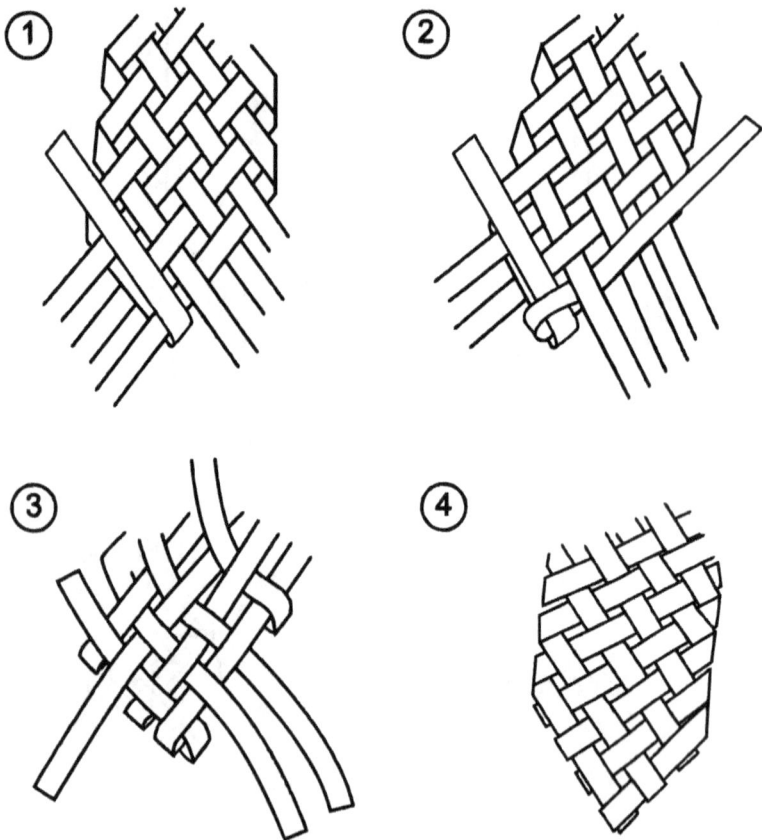

Whipping the Ends

A-46. The raw, cut end of a rope has a tendency to untwist and should always be knotted or fastened in some manner. Whipping is one method of fastening the end of a rope. A rope is whipped by wrapping the end tightly with a small cord. This method is particularly satisfactory because it does not greatly increase the size of the rope. The whipped end of a rope will still thread through blocks or other openings. Before cutting a rope, place two whippings on the rope 1~2 inches apart and make the cut between the whippings. This will prevent the cut ends from untwisting immediately after they are cut.

A-47. Make a bight near one end of the small whip-cord and lay both ends of the small cord along one side of the rope. The bight should project beyond the end of the rope about one-half inch. Wrap the running end of the cord tightly around the rope and cord starting at the end of the whipping (this will be farthest from the end of the rope). The wrap should be in the same direction as the twist of the rope strands. Continue wrapping the cord around the rope, keeping it tight, to within about one-half inch of the end.

A-48. At this point, slip the running end through the bight of the cord. The standing part of the cord can then be pulled until the bight of the cord is pulled under the whipping and the cord is tightened. Cut the ends of the cord at the edge of the whipping, leaving the rope end whipped (see Figure A-25 on page 304).

Figure A-25. Whipping the End of a Rope

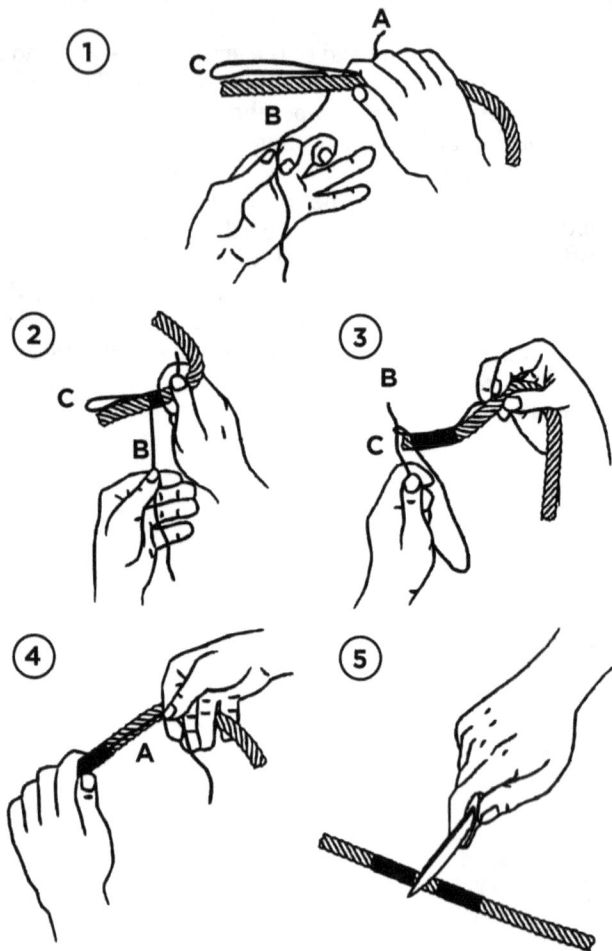

GLOSSARY

SECTION 1: ACRONYMS AND ABBREVIATIONS

- **ATP**: Army tactics publication
- **CAT**: combat application tourniquet
- **COLDER**: clean, overheating, loose layers, dry, examine, and repair
- **DA**: Department of the Army
- **EPA**: Evasion plan of action
- **EVC**: Evasion chart
- **FM**: Field manual
- **GPS**: Global Positioning System
- **IFAK**: Improved first aid kit
- **ISG**: Isolated Soldier guidance
- **IV**: intravenous
- **JP**: Joint publication
- **SERE**: Survival, evasion, resistance, and escape
- **TC**: Training circular
- **U.S.**: United States

SECTION 2: TERMS

- **Army personnel recovery**: the military efforts taken to prepare for and execute the recovery and reintegration of isolated personnel.
- **Planning**: the art and science of understanding a situation, envisioning a desired future, and laying out effective ways of bringing that future about.
- **Preparation**: those activities performed by units and Soldiers to improve their ability to execute an operation.
- **Survival, evasion, resistance, and escape**: actions performed by isolated personnel designed to ensure their health, mobility, safety, and honor in anticipation of or preparation for their return to friendly control.

REFERENCES

RELATED PUBLICATIONS

- **ADP 1-02**. Terms and Military Symbols. 14 August 2018.
- **DOD Dictionary of Military and Associated Terms**. June 2018.
- **JP 3-50**. Personnel Recovery. 2 October 2015.
- **ADP 5-0**. The Operations Process. 17 May 2012.
- **ATP 3-50.20**. Survival, Evasion, Resistance, and Escape (SERE) Preparation and Training. 29 November 2017.
- **ATP 3-50.22**. Evasion and Recovery. 28 November 2017.
- **FM 3-05.70**. Survival. 17 May 2002.[1]
- **FM 3-50**. Army Personnel Recovery. 2 September 2014.
- **FM 27-10**. The Law of Land Warfare. 18 July 1956.
- **STP 21-1-SMCT**. Soldier's Manual of Common Tasks Warrior Skills Level 1, 28 September 2017.
- **TC 3-25.26**. Map Reading and Land Navigation. 15 November 2013.[2]
- **TM 3-34.86/MCRP 3-17.7J**. Rigging Techniques, Procedures, and Applications. 16 July 2012.

---

1. Also available from Carlile Media: ISBN 1547209461.
2. Also available from Carlile Media: ISBN 1977649270.

INDEX

Entries are listed by Paragraph number.

A

Azimuth, 7-39, 7-51, 7-54~7-59, 7-67~7-68, 7-70, 7-81, 7-85

B

Bandages, 2-6, 2-9, 2-24, 2-47
    sterile, 2-47
Bleeding, 1-13
    from cuts, 1-13
Boiling, 2-5, 2-24, 2-66, 2-78, 2-87, 2-163~2-164, 3-34~3-35, 3-39~3-40, 4-27~4-31, 4-48, 4-71, 4-128, 4-137, 4-167, 5-1, 8-3
Bowline, A-12~A-16, A-26
    French, A-15~A-16
Bowline on a Bight, A-14
Braiding techniques, A-42~A-45
    three-plait braid, A-43
    broad braid, A-44~A-45
Brain injury, 2-56~2-58
Burns, 2-41, 2-43~2-47, 2-77, 2-79, 2-82, 2-84, 2-145, 2-154, 4-1, 5-3, 5-17, 5-20~5-21, 5-24, 5-27
    to the face, 2-43
Butter, 2-46, 4-95, 4-99

C

Cardinal direction, determining 7-22, 7-87
    shadow tip method, 7-22, 7-24~7-25
    equal-shadow method, 7-25

watch method, 7-26~7-29
24-hour clock method, 7-30
using the moon, 7-31
northern sky, 7-33
stars, 7-32, 7-34~7-36
southern sky, 7-37~7-39
improvised compass, 7-40~7-42
floating needle or leaf compass, 7-43
other means of determining direction, 7-44~7-46

Care under fire, 2-1

Catching features, 7-66

Clubs, 8-10~8-14
simple club, 8-11
sling club, 8-14
weighted, 8-12~8-13

Combat pill kit, 2-1

Compass, 1-8, 7-40, 7-42, 7-49~7-50, 7-52, 7-54~7-55, 7-58~7-59, 7-65, 7-68, 7-70~7-72
improvised, 7-42

Cooking, 2-156, 4-23~4-28, 4-30, 4-32, 4-36, 4-38, 4-43, 4-71, 4-113, 4-130~4-131, 4-136, 4-139, 4-141~4-142, 4-149, 4-164, 5-28~5-29, 6-27, 8-42, 8-44, 8-46~8-47, 8-51

Cooking and eating utensils, 8-43

D

Dead reckoning, 7-47, 7-50~7-53, 7-55~7-56, 7-60~7-61, 7-83
Digital ligation, 2-20, 2-35
Dressings, 2-6, 2-24, 2-43, 2-53

E

Elevation, 2-20, 2-25, 2-35, 7-63
Evasion plan of action, 1-3, 7-1

F

Field-expedient weapons, 8-7
Flail chest, 2-37
Flotation devices, 7-19~7-20

Fraps, A-2

G

GPS, 1-8, 7-47~7-49, 8-1
Grease, 2-46, 5-47, 6-7

H

Handrails, 7-64~7-65
Head injury, 2-51, 7-1
Hitch, 4-155, 6-16, 6-22, 6-43, A-2, A-4~A-5, A-19~A-25, A-28~A-30
    clove hitch, A-24
    half hitch, 6-16, 6-22, A-19~A-23
    Prusik, A-26
    timber hitch, A-22

I, J, K,

Ice, 3-13

L

Lashings, A-27~A-30
    square lashing, A-28
    diagonal lashing, A-29
    shear lashing, A-30
Leaching, 4-26, 4-72
Lubricants and glue, 8-3~8-6

M

Maintaining equipment, 8-1~8-6
Map orientation, 7-72~7-73
MARCH (acronym), 2-1~2-2, 2-7~2-8, 2-44, 2-110, 7-1
Measuring azimuth, 7-81
Measuring distances, 7-77
    graphic scales, 7-78
    ground distance, straight line, 7-79

ground distance, curved line, 7-80
plotting a direct line, 7-82~7-83

Mild brain injury, 2-54

Mitigation, 1-5, 2-3, A-34

Movement, 1-8, 1-21~1-22, 2-3, 2-36, 2-58, 2-104, 2-107, 2-136, 2-144, 2-153, 3-35, 4-83, 5-33, 6-16, 7-1~7-7, 7-28, 7-49, 7-52~7-53, 7-56, 7-60~7-63, 7-65, 7-83, 7-85~7-87 jungle, 7-7

N

Navigation, 1-3, 1-8~1-9, 1-19, 1-23, 7-39, 7-47~7-49, 7-52, 7-67~7-69, 7-83, 7-85
at night, 7-85

Navigation methods, 7-47~7-60
Global Positioning System, 7-48~7-49
dead reckoning, 7-50~7-60
terrain association, 7-61~7-62

Navigational attack points, 7-67

O

Operations, 1-6, 2-1, 2-6, 2-41, 2-95, 3-43, 4-46, 7-1

P, Q

Pain, 2-3, 2-36, 2-43, 2-45~2-47, 2-54, 2-59, 2-70~2-71, 2-79~2-80, 2-82, 2-101, 2-103, 2-107, 2-109, 2-113, 2-117, 2-133, 2-148, 2-150, 2-157, 3-36, 4-147
neck, 2-54

Planning, 1-5~1-8, 4-96, 4-98, 7-85

Pneumothorax, 2-1

Pottery, 8-47~8-48, 8-51

Preparation, 1-5~1-8, 1-18, 2-75, 2-78, 2-80, 2-82~2-83, 2-88, 4-22, 4-35, 4-130, 4-136, 5-6, 5-46, 5-54, 6-11

Pressure points, 2-26~2-27, 2-32

Prevention, 1-5, 2-3, 2-97, 2-103, 2-134, 2-165, 7-2

Protection, 1-3, 1-7, 1-19, 1-23, 2-103, 4-44, 6-2, 6-8~6-9, 6-13, 6-16, 6-29, 8-9, 8-53, A-34

Psychology of survival, 1-9~1-18

R

Rafts, 7-14~7-18
    brush, 7-14, 7-15
    Australian poncho raft, 7-14, 7-16~7-17
    log, 7-18~7-19
    two-log, 7-19
Recognition, 1-5, 2-3
Rope construction, A-31~A-48
Rope terminology, A-1~A-2
Rucksack construction, 8-42
    horseshoe pack, 8-42
    square pack, 8-42

S

Serious brain injury, 2-55~2-56
Sheepshank, A-19, A-25
Shelter, 1-7, 1-9, 1-19, 2-5, 2-61, 2-85, 2-163, 4-41, 4-147, 5-5, 5-7, 6-1, 6-8~6-16, 6-19~6-25, 6-27~6-32, 6-35~6-36, 6-38~6-44, 7-1, 7-4, 7-6, 8-2, A-27
    natural shelter construction, 6-23~6-38
    debris hut, 6-38
    desert shelters, 6-35
    field-expedient lean-to, 6-31~6-34
    raised platforms shelter, 6-29~6-30
    snow caves, 6-25~6-27
    swamp bed, 6-37
    tree pit shelter, 6-28
Shock, 2-1, 2-9, 2-43~2-44, 2-48, 2-51, 2-60~2-62, 2-72, 2-114, 2-132~2-133, 2-153, 7-1
Situational understanding, 1-21~1-23
Soldier guidance, 1-3, 7-1
Speir knot, A-12, A-17~A-18
Square Lashing, A-2, A-27~A-28
Stress, 1-6, 1-10~1-15, 4-81
Survival psychology, 1-9~1-18
Survival medicine, 1-3~1-5, 1-19~1-20, 1-23, 2-1~2-3, 2-6, 2-76~2-77, 2-80, 2-82~2-84, 2-113, 2-158
Sustenance, 1-3, 1-6, 1-15, 1-19, 1-23

T

Terrain features, 7-74~7-76
    hills, 6-14, 7-74
    saddles, 7-74
    valleys, 7-74
    ridges, 7-58, 7-74
    depressions, 3-2, 6-14, 7-74
    draws, 7-75
    spurs, 7-75
    cliffs, 3-33, 7-75
    cuts, 7-76
    fills, 7-76

Training, 1-3, 1-8, 1-18, 4-62

Treatment, 1-5, 2-1, 2-7~2-8, 2-41, 2-64, 2-66~2-68, 2-70, 2-74~2-75, 2-79, 2-81~2-84, 2-96~2-97, 2-99, 2-103, 2-109~2-110, 2-113~2-115, 2-125, 2-131, 2-133, 2-135, 2-137~2-138, 2-140~2-142, 2-158, 2-166, 3-43
    sequence of, 2-1

U

Utensils, 2-5, 2-163~2-164, 5-29, 8-43

V

Valley, 4-83, 7-74~7-75

W, X

Water, 2-38
    ice, 2-46~2-47, 2-54, 2-103, 2-108, 3-7, 3-12~3-13, 6-19, 7-5, 7-83
Water crossings, 7-8~7-20
    rapids, 7-11~7-13
    rivers and streams, 7-9~7-10
Weapons, 2-50, 2-57, 4-81, 7-16, 8-9, 8-14, 8-19~8-21, 8-25
Weaving, 8-53~8-55

Y, Z

Yarrow, 2-80, 2-82, 2-84